Y0-BLH-489

MODERN ALGEBRA

Reuben Sandler

L. Sheila Foster
California State University at Long Beach

HARPER & ROW, PUBLISHERS
New York   Hagerstown   San Francisco   London

To Dusty

Sponsoring Editor: Charlie Dresser
Project Editor: Brigitte Pelner
Designer: Gayle Jaeger
Production Supervisor: Marion A. Palen
Compositor: Syntax International Pte. Ltd.
Printer and Binder: The Maple Press Company

Modern Algebra
Copyright © 1978 by Reuben Sandler and L. Sheila Foster

All rights reserved. Printed in the United States of America. No part of this book may be used or reproduced in any manner whatsoever without written permission except in the case of brief quotations embodied in critical articles and reviews. For information address Harper & Row, Publishers, Inc., 10 East 53rd Street, New York, N.Y. 10022.

Library of Congress Cataloging in Publication Data

Sandler, Reuben.
    Modern algebra.

  1. Algebra, Abstract. I. Foster, Leslie
Sheila, Date–   joint author. II. Title.
QA162.S26   512'.02   77-25263
ISBN 0-06-045718-X

CONTENTS

PREFACE v

PART I. GROUP THEORY 1

1. EXAMPLES OF GROUPS 3
2. THE DEFINITION OF A GROUP 12
3. SOME IMPORTANT GROUPS 18
   Groups of integers modulo n 20
   Groups arising in geometrical contexts 22
   Permutation groups 24
4. DEVELOPMENT OF ELEMENTARY GROUP PROPERTIES 28
5. SUBGROUPS AND CYCLIC SUBGROUPS 37
   Subgroups 37
   Cyclic subgroups 43
   Division algorithm 46
6. EQUIVALENCE RELATIONS, COSETS, NORMAL SUBGROUPS, AND LAGRANGE'S THEOREM 53
   Equivalence relations 53
   Cosets 62
   Normal subgroups 66
   Lagrange's theorem 71
7. ISOMORPHISMS, AUTOMORPHISMS, HOMOMORPHISMS 75
   Isomorphisms 75

	Automorphisms 81
	Homomorphisms 86
8	THE STRUCTURE OF FINITE GROUPS 95
	Finite abelian groups 95
	Centers, centralizers, and normalizers 99
	Conjugacy 104
	Proofs of Cauchy's theorems and Sylow's theorem 108
9	PERMUTATION GROUPS 111

PART II. RING THEORY   131

10	EXAMPLES AND AXIOMS 133
11	ELEMENTARY RING THEORY 140
	Basic concepts 140
	Subgroups and ideals 143
	Homomorphisms 150
	Characteristic of a ring 156
12	SPECIAL CLASSES OF RINGS 161
	Euclidean rings 161
	Polynomial rings 170

PART III. FIELD THEORY   181

13	EXAMPLES AND AXIOMS 183
14	SUBFIELDS AND EXTENSION FIELDS 192
15	ROOTS OF POLYNOMIALS IN $F[x]$ 199
16	GALOIS THEORY 207

PREFACE

An introductory course in modern algebra is included in the undergraduate curriculum of most students of mathematics and physics, as well as many students of economics. Increasing numbers of students in other fields are also studying the subject. The present book is intended to service such a course of either a one-semester or two-quarter duration. It is flexible enough to work as a text for most introductory courses which do not include linear algebra. (For a combined course including linear algebra, another text would be required in combination with this one.) Part I on group theory could fill up almost a whole semester of a course oriented in that direction. On the other hand, the introductory chapters of the parts on group theory, ring theory, and field theory can be combined to give a more diversified course. With the exception of the chapter on Galois theory, none of the material on rings and fields depends on the advanced material on groups (Chapters 8 and 9).

The authors have attempted to write a book for the intelligent student—a book which is more helpful than the usual texts. Among the methods we have employed are thorough investigations of many familiar and unfamiliar examples; the illustration of each new concept with examples; questions to the student which indicate the logical connections which he/she should make at that point in the text (this helps to test comprehension as the student progresses page by page); good exercises inserted at appropriate points in the text as an aid to understanding the material; careful discussion of the aims of each section; enough of an overall discussion to make parts of the subject band together in a unified whole for the student.

Among the many people whom we would like to thank, special mention goes to Harry D. Eylar of California State University, Long Beach, for his critical reading and technical assistance, Betty Nosko for her expert typing under pressure, and especially to William D. Foster and Dale Herron for their help in every imaginable way.

GROUP THEORY

1 Examples of groups

We shall begin our study of groups by looking closely at many examples. We will not define the term "group" yet, but the study of a number of examples will lead to the definition in the next chapter. For now, we will be looking at some familiar and unfamiliar mathematical systems and studying some of their properties.

In algebra we are concerned not only with sets, but also with operations like addition and multiplication defined on the sets. An **algebraic system** consists of a set together with one or more operations defined on it. These terms will be defined rigorously in Chapter 2.

EXAMPLE 1.1

Our first example will deal with the set of real numbers, $\mathbb{R}$. Two operations on the real numbers are addition and multiplication. Initially, we will consider the operation of addition alone. We will call the system—real numbers and addition—$(\mathbb{R}, +)$. For any two real numbers, x and y, there is a third real number, given by $x + y$. There are many properties of the system $(\mathbb{R}, +)$ which we could study, but at this time we are only interested in a few specific ones.

For example, there is a real number, 0, the additive identity, which has the property that

$$0 + x = x + 0 = x, \quad \text{for all } x \in \mathbb{R}.$$

That is, adding 0 to any real number does not change it.

Also, every real number has an **additive inverse** with respect to 0. That is, for any x,

$$x + (-x) = (-x) + x = 0.$$

The number $(-x)$ is called the additive inverse of x. For example, $1+(-1)=(-1)+1=0$, or -1 is the additive inverse of 1; $(-3\frac{1}{2})+(-(-3\frac{1}{2}))=-(-3\frac{1}{2})+(-3\frac{1}{2})=0$ or $-(-3\frac{1}{2})$ is the additive inverse of $(-3\frac{1}{2})$.

Finally, let us remember that addition of real numbers is **associative**, or for any real numbers x, y, and z,

$$x+(y+z)=(x+y)+z.$$

To conclude, in the system consisting of the real numbers and addition, the properties with which we are the most concerned are those of existence of an identity, existence of inverses, and associativity.

EXAMPLE 1.2

Our next example also deals with the real numbers, but uses only the operation of multiplication. Besides switching from addition to multiplication, we will omit the real number 0. Let us look at the system $(\mathbb{R}-\{0\},\cdot)$, the nonzero real numbers and the operation of multiplication. For any two nonzero real numbers x and y, the product $x\cdot y$ is also a nonzero real number. (Why is this true?)

There is a nonzero real number 1 which has the property

$$1\cdot x=x\cdot 1=x, \quad \text{for } x \text{ any nonzero real number.}$$

That is, no nonzero real number is changed by multiplication by 1.

Also, for every nonzero real number, there is a **multiplicative inverse** with respect to 1. For any x, $x\neq 0$, there is a number $1/x$ with the property

$$x\cdot\frac{1}{x}=\frac{1}{x}\cdot x=1.$$

The number $1/x$ is called the multiplicative inverse of x. For example,

$$-\frac{3}{2}\cdot\frac{1}{-\frac{3}{2}}=\frac{1}{-\frac{3}{2}}\cdot-\frac{3}{2}=1$$

or $1/-\frac{3}{2}$ is the multiplicative inverse of $-\frac{3}{2}$.

Finally, recall that the multiplication of real numbers is associative:

$$(x\cdot y)\cdot z=x\cdot(y\cdot z).$$

Again, the properties of the system $(\mathbb{R}-\{0\},\cdot)$ with which we are most concerned are those of existence of an identity, existence of inverses, and associativity.

EXAMPLE 1.3
The system $(Z, +)$ is composed of the set of integers, $Z = \{0, \pm 1, \pm 2, \ldots\}$, and the operation of addition. The sum of any two integers is again an integer (why?), and the operation of addition on the integers is associative. 0 is an integer and for any integer n,

$$n + 0 = 0 + n = n$$

so that 0 is the additive identity. Also, for every integer n, there is an integer $-n$ with the property that

$$n + (-n) = (-n) + n = 0.$$

The same three properties hold for this set.

EXAMPLE 1.4
This example should also seem familiar but will show that the three properties we have discussed are not universally held by all algebraic systems. Consider the nonzero integers under multiplication, $(Z - \{0\}, \cdot)$. The product of any two integers is always an integer (why?), 1 is a nonzero integer and $1 \cdot n = n \cdot 1 = n$ for all integers n, and multiplication of integers is associative. But what about multiplicative inverses? Is there an *integer* x such that $3 \cdot x = x \cdot 3 = 1$? No, for the only solution to $3 \cdot x = 1$ is $x = \frac{1}{3}$, and $\frac{1}{3}$ *is not an integer*. For this system, only three of the four properties hold.

EXERCISES
1. Suppose we try to amend Example 1.4 by taking the nonzero integers and including all solutions of the form x where $n \cdot x = 1$. Do we end up with a system in which every element has a multiplicative inverse? Does this new system satisfy the four properties discussed? If not, can more elements be included so that the system does satisfy all the properties? If so, do you recognize the resultant system?
2. Can you show that the set of *even* integers satisfies the rules we have been discussing relative to the operation of addition? What about the *odd* integers?

The examples which we will investigate next are different in structure from the previous ones; they deal with the study of matrices.

EXAMPLE 1.5
Let $M_{2 \times 3}$ be the set of *all* 2×3 matrices whose entries are real numbers,

$$M_{2 \times 3} = \left\{ \begin{pmatrix} x_{11} & x_{12} & x_{13} \\ x_{21} & x_{22} & x_{23} \end{pmatrix} \middle| \text{each } x_{ij} \text{ is a real number} \right\}.$$

If

$$\begin{pmatrix} x_{11} & x_{12} & x_{13} \\ x_{21} & x_{22} & x_{23} \end{pmatrix} \text{ and } \begin{pmatrix} y_{11} & y_{12} & y_{13} \\ y_{21} & y_{22} & y_{23} \end{pmatrix}$$

are two elements in $M_{2 \times 3}$, addition is defined by

$$\begin{pmatrix} x_{11} & x_{12} & x_{13} \\ x_{21} & x_{22} & x_{23} \end{pmatrix} + \begin{pmatrix} y_{11} & y_{12} & y_{13} \\ y_{21} & y_{22} & y_{23} \end{pmatrix}$$
$$= \begin{pmatrix} x_{11} + y_{11} & x_{12} + y_{12} & x_{13} + y_{13} \\ x_{21} + y_{21} & x_{22} + y_{22} & x_{23} + y_{23} \end{pmatrix}.$$

Since the result is also a 2 × 3 matrix, we can see that the sum of any two elements of $M_{2 \times 3}$ is also in $M_{2 \times 3}$.

The matrix

$$0_{2 \times 3} = \begin{pmatrix} 0 & 0 & 0 \\ 0 & 0 & 0 \end{pmatrix}$$

is a member of $M_{2 \times 3}$ and acts as an additive identity since

$$\begin{pmatrix} x_{11} & x_{12} & x_{13} \\ x_{21} & x_{22} & x_{23} \end{pmatrix} + \begin{pmatrix} 0 & 0 & 0 \\ 0 & 0 & 0 \end{pmatrix} = \begin{pmatrix} 0 & 0 & 0 \\ 0 & 0 & 0 \end{pmatrix} + \begin{pmatrix} x_{11} & x_{12} & x_{13} \\ x_{21} & x_{22} & x_{23} \end{pmatrix}$$
$$= \begin{pmatrix} x_{11} & x_{12} & x_{13} \\ x_{21} & x_{22} & x_{23} \end{pmatrix}$$

for any element $\begin{pmatrix} x_{11} & x_{12} & x_{13} \\ x_{21} & x_{22} & x_{23} \end{pmatrix}$ in $M_{2 \times 3}$.

For $M = \begin{pmatrix} x_{11} & x_{12} & x_{13} \\ x_{21} & x_{22} & x_{23} \end{pmatrix} \in M_{2 \times 3}$, we define

$$-M = \begin{pmatrix} -x_{11} & -x_{12} & -x_{13} \\ -x_{21} & -x_{22} & -x_{23} \end{pmatrix}.$$

$-M$ is also a member of $M_{2 \times 3}$ and acts as an additive inverse of M with respect to $0_{2 \times 3}$ since

$$M + (-M) = (-M) + M = 0_{2 \times 3}.$$

Finally, it can be shown that for any three matrices, M, N, and K, in $M_{2 \times 3}$,

$$M + (N + K) = (M + N) + K \quad \text{(why?)};$$

therefore, the associative law holds.

Examples of groups

Reflecting on these properties, we can see that they hold for sets of matrices of any size, 1×4, 7×13, and so on. There is no magic about $M_{2 \times 3}$, we just used it as an example of a whole class of sets of matrices. What we proved for $(M_{2 \times 3}, +)$ could just as easily be demonstrated for the system $(M_{n \times m}, +)$ where $M_{n \times m}$ is the set of all $n \times m$ matrices.

EXERCISES
1. Prove that $(M_{2 \times 3}, +)$ is associative.
2. a. Using the methods of this chapter define $(M_{1 \times 4}, +)$ and show that it has the properties shown for $(M_{2 \times 3}, +)$.
   b. How would the general case, $(M_{n \times m}, +)$, be handled?
3. What would $(M_{1 \times 1}, +)$ be? What other system that we have studied is it similar to?

Our next example also deals with a set of matrices but with an operation other than addition. The example uses the operation of matrix multiplication and leads to some interesting computations.

EXAMPLE 1.6
Consider the set of all 2×2 matrices,

$$M_{2 \times 2} = \left\{ \begin{pmatrix} x_{11} & x_{12} \\ x_{21} & x_{22} \end{pmatrix} \middle| x_{11}, x_{12}, x_{21}, x_{22} \text{ are real numbers} \right\}.$$

If $M = \begin{pmatrix} x_{11} & x_{12} \\ x_{21} & x_{22} \end{pmatrix}$ and $N = \begin{pmatrix} y_{11} & y_{12} \\ y_{21} & y_{22} \end{pmatrix}$ we define the matrix product, $M \cdot N$, by the following rule:

$$M \cdot N = \begin{pmatrix} x_{11} & x_{12} \\ x_{21} & x_{22} \end{pmatrix} \cdot \begin{pmatrix} y_{11} & y_{12} \\ y_{21} & y_{22} \end{pmatrix}$$

$$= \begin{pmatrix} x_{11}y_{11} + x_{12}y_{21} & x_{11}y_{12} + x_{12}y_{22} \\ x_{21}y_{11} + x_{22}y_{21} & x_{21}y_{12} + x_{22}y_{22} \end{pmatrix}.$$

A specific example follows: If $M = \begin{pmatrix} 1 & 2 \\ 3 & 4 \end{pmatrix}$ and $N = \begin{pmatrix} -2 & 4 \\ -3 & 2 \end{pmatrix}$, then

$$M \cdot N = \begin{pmatrix} 1 & 2 \\ 3 & 4 \end{pmatrix} \begin{pmatrix} -2 & 4 \\ -3 & 2 \end{pmatrix} = \begin{pmatrix} 1(-2) + 2(-3) & 1 \cdot 4 + 2 \cdot 2 \\ 3(-2) + 4(-3) & 3 \cdot 4 + 4 \cdot 2 \end{pmatrix}$$

$$= \begin{pmatrix} -8 & 8 \\ -18 & 20 \end{pmatrix}.$$

$(M_{2 \times 2}, \cdot)$ is not exactly the system we want to investigate. It is too large;

we will explore the reasons for this in the exercises. We restrict ourselves, instead, to a subset of $M_{2\times 2}$.

Consider the system $(N_{2\times 2}, \cdot)$ where $N_{2\times 2}$ consists of those 2×2 matrices whose determinants are not zero. (The determinant of the 2×2 matrix $\begin{pmatrix} x_{11} & x_{12} \\ x_{21} & x_{22} \end{pmatrix}$ is $x_{11}x_{22} - x_{12}x_{21}$. For example, $\begin{pmatrix} 1 & 2 \\ 3 & 4 \end{pmatrix}$ is in $N_{2\times 2}$ since $\det \begin{pmatrix} 1 & 2 \\ 3 & 4 \end{pmatrix} = 1 \cdot 4 - 2 \cdot 3 = -2 \neq 0$, but $\begin{pmatrix} 1 & -2 \\ -3 & 6 \end{pmatrix}$ is not in $N_{2\times 2}$ since $\det \begin{pmatrix} 1 & -2 \\ -3 & 6 \end{pmatrix} = 1 \cdot 6 - (-2) \cdot (-3) = 0$.

First, if $M \in N_{2\times 2}$ and $N \in N_{2\times 2}$, is $M \cdot N$ in $N_{2\times 2}$? Let us try an example. If $M_1 = \begin{pmatrix} -1 & 2 \\ 3 & 4 \end{pmatrix}$ and $N_1 = \begin{pmatrix} 2 & 1 \\ 3 & 5 \end{pmatrix}$, M_1 and N_1 belong to $N_{2\times 2}$ since $\det M_1 = -10$ and $\det N_1 = 7$. $M_1 \cdot N_1 = \begin{pmatrix} -1 & 2 \\ 3 & 4 \end{pmatrix}\begin{pmatrix} 2 & 1 \\ 3 & 5 \end{pmatrix} = \begin{pmatrix} 4 & 9 \\ 18 & 23 \end{pmatrix}$ and $M_1 \cdot N_1$ belongs to $N_{2\times 2}$ since $\det \begin{pmatrix} 4 & 9 \\ 18 & 23 \end{pmatrix} = -70$.

Since we cannot prove a result by doing specific examples, we have to attack the general case. Let $M = \begin{pmatrix} x_{11} & x_{12} \\ x_{21} & x_{22} \end{pmatrix}$ and $N = \begin{pmatrix} y_{11} & y_{12} \\ y_{21} & y_{22} \end{pmatrix}$. We know only that $\det M = x_{11}x_{22} - x_{12}x_{21} \neq 0$ and $\det N = y_{11}y_{22} - y_{12}y_{21} \neq 0$.

$$M \cdot N = \begin{pmatrix} x_{11}y_{11} + x_{12}y_{21} & x_{11}y_{12} + x_{12}y_{22} \\ x_{21}y_{11} + x_{22}y_{21} & x_{21}y_{12} + x_{22}y_{22} \end{pmatrix}$$

which is a 2×2 matrix. We still need to verify that the determinant of the product is not zero.

$\det M \cdot N = (x_{11}y_{11} + x_{12}y_{21})(x_{21}y_{12} + x_{22}y_{22})$
$\qquad - (x_{11}y_{12} + x_{12}y_{22})(x_{21}y_{11} + x_{22}y_{21}).$

Expanding $\det M \cdot N$ and manipulating it algebraically, we get

$\det M \cdot N = \cancel{x_{11}y_{11}x_{21}y_{12}} + x_{11}y_{11}x_{22}y_{22} + x_{12}y_{21}x_{21}y_{12} + \cancel{x_{12}y_{21}x_{22}y_{22}}$
$\qquad - \cancel{x_{11}y_{12}x_{21}y_{11}} - x_{11}y_{12}x_{22}y_{21} - x_{12}y_{22}x_{21}y_{11}$
$\qquad - \cancel{x_{12}y_{22}x_{22}y_{21}}$
$\qquad = x_{11}y_{11}x_{22}y_{22} - x_{11}y_{12}x_{22}y_{21} + x_{12}y_{21}x_{21}y_{12} - x_{12}y_{22}x_{21}y_{11}$
$\qquad = x_{11}x_{22}(y_{11}y_{22} - y_{12}y_{21}) - x_{12}x_{21}(y_{11}y_{22} - y_{12}y_{21})$
$\qquad = (x_{11}x_{22} - x_{12}x_{21})(y_{11}y_{22} - y_{12}y_{21})$
$\qquad = (\det M)(\det N).$

Examples of groups

But det $M \neq 0$ and det $N \neq 0$ so that (det M)(det N) $\neq 0$ (why?), and we can conclude that if M and N are both in $N_{2 \times 2}$, so is the product $M \cdot N$.

Next, define $I_{2 \times 2} = \begin{pmatrix} 1 & 0 \\ 0 & 1 \end{pmatrix}$. For any $M = \begin{pmatrix} x_{11} & x_{12} \\ x_{21} & x_{22} \end{pmatrix}$ in $N_{2 \times 2}$,

$$M \cdot I_{22} = \begin{pmatrix} x_{11} & x_{12} \\ x_{21} & x_{22} \end{pmatrix} \begin{pmatrix} 1 & 0 \\ 0 & 1 \end{pmatrix} = \begin{pmatrix} x_{11} & x_{12} \\ x_{21} & x_{22} \end{pmatrix} = M$$

and

$$I_{2 \times 2} \cdot M = \begin{pmatrix} 1 & 0 \\ 0 & 1 \end{pmatrix} \begin{pmatrix} x_{11} & x_{12} \\ x_{21} & x_{22} \end{pmatrix} = \begin{pmatrix} x_{11} & x_{12} \\ x_{21} & x_{22} \end{pmatrix} = M$$

so that $I_{2 \times 2}$ acts as an identity element for the system $(N_{2 \times 2}, \cdot)$.

Let $M = \begin{pmatrix} x_{11} & x_{12} \\ x_{21} & x_{22} \end{pmatrix}$, det $M \neq 0$ ($M \in N_{2 \times 2}$). Define

$$M^{-1} = \begin{pmatrix} \dfrac{x_{22}}{\det M} & \dfrac{-x_{12}}{\det M} \\ \dfrac{-x_{21}}{\det M} & \dfrac{x_{11}}{\det M} \end{pmatrix}.$$

For a specific example, if $M = \begin{pmatrix} 2 & 1 \\ 4 & 3 \end{pmatrix}$, det $M = 2$, and thus

$$M^{-1} = \begin{pmatrix} \frac{3}{2} & -\frac{1}{2} \\ -2 & \frac{2}{2} \end{pmatrix}.$$

In our example, we can see that

$$M \cdot M^{-1} = \begin{pmatrix} 2 & 1 \\ 4 & 3 \end{pmatrix} \cdot \begin{pmatrix} \frac{3}{2} & -\frac{1}{2} \\ -2 & \frac{2}{2} \end{pmatrix} = \begin{pmatrix} 1 & 0 \\ 0 & 1 \end{pmatrix} = M^{-1} \cdot M$$

and M^{-1} is an inverse for M.

This same result can be shown for the general matrix $M = \begin{pmatrix} x_{11} & x_{12} \\ x_{21} & x_{22} \end{pmatrix}$ and the corresponding M^{-1}. Do the multiplication and show in general that

$$M \cdot M^{-1} = M^{-1} \cdot M = I_{2 \times 2}.$$

Our last concern with $(N_{2 \times 2}, \cdot)$ is to show that the system is associative. This is a purely computational exercise.

An interesting thing to notice about both $(N_{2 \times 2}, \cdot)$ and $(M_{2 \times 2}, \cdot)$ and, in fact, any matrix multiplication is that it is not commutative. That is, it is not necessarily true that $M \cdot N = N \cdot M$.

For example,

$$\begin{pmatrix} 1 & 2 \\ 3 & 4 \end{pmatrix} \begin{pmatrix} 1 & 1 \\ 3 & 5 \end{pmatrix} = \begin{pmatrix} 7 & 11 \\ 15 & 23 \end{pmatrix}$$

although

$$\begin{pmatrix} 1 & 1 \\ 3 & 5 \end{pmatrix} \begin{pmatrix} 1 & 2 \\ 3 & 4 \end{pmatrix} = \begin{pmatrix} 4 & 6 \\ 18 & 26 \end{pmatrix}.$$

How many of the other systems studied in this chapter are commutative?

EXERCISES

1. If $M = \begin{pmatrix} x_{11} & x_{12} \\ x_{21} & x_{22} \end{pmatrix}$ with det $M \neq 0$, show that $M \cdot M^{-1} = M^{-1} \cdot M = I_{2 \times 2}$ where M^{-1} is as defined in this chapter. (This proves that every element of $N_{2 \times 2}$ has an inverse relative to $I_{2 \times 2}$.)
2. Prove that any 2 × 2 matrix with zero determinant cannot have an inverse. That is, if $\det(M) = 0$, then there is no matrix N with the property that $M \cdot N = I_{2 \times 2}$.
3. Which of the other systems that we have studied have the commutative property (i.e., have the property that $a \circ b = b \circ a$ where a and b are any two elements of the set and $\circ$ is the operation)?
4. Let $\mathscr{C}_{2 \times 2} = \left\{ \begin{pmatrix} a & 0 \\ 0 & a \end{pmatrix} \middle| a \text{ is a nonzero real number} \right\}$ and consider the system $(\mathscr{C}_{2 \times 2}, \cdot)$.
   a. Show that the product of any two elements in $\mathscr{C}_{2 \times 2}$ is an element of $\mathscr{C}_{2 \times 2}$.
   b. Show that $I_{2 \times 2} \in \mathscr{C}_{2 \times 2}$ and acts as an identity for $\mathscr{C}_{2 \times 2}$.
   c. Determine whether elements of $\mathscr{C}_{2 \times 2}$ have inverses in $\mathscr{C}_{2 \times 2}$. Prove your results.
   d. Is the system $(\mathscr{C}_{2 \times 2}, \cdot)$ commutative?

Our last example in this chapter deals with a set of functions. The techniques are somewhat different; the basic concept used is that a function is determined by what it does to every point of its domain.

EXAMPLE 1.7

Let $\mathscr{F}$ be the set of *all* functions from the real numbers, $\mathbb{R}$, to the real numbers. Then a function $f \in \mathscr{F}$ is defined by its action on every real number; knowing all pairs $(x, f(x))$ determines *exactly* what f is.

The operation we will work with is addition of functions. (There are other operations on functions like composition and multiplication which may be considered later.) If f and g are two functions, then the function

$f + g$ is defined by

$(f + g)(x) = f(x) + g(x)$ for all real numbers x.

For any real number x, we know what $(f + g)(x)$ is, and therefore we know what the *function* $f + g$ is. $f + g$ is a function on the real numbers whenever f and g are (why?).

Also, if f, g, and h are functions, then

$f + (g + h) = (f + g) + h$.

To see this, ask what each of the functions $f + (g + h)$ and $(f + g) + h$ is. What do they do to a typical real number, x? Are they always the same for each x? If so, then they are the *same function*, and it follows that the addition of functions is associative. Do the necessary calculations and answer the questions to verify this.

Next, there is a function $0 \in \mathscr{F}$, defined by $0(x) = 0$ for all $x \in \mathbb{R}$. This function satisfies

$0 + f = f + 0 = f$ for any function $f \in \mathscr{F}$

because $(f + 0)(x) = f(x) + 0(x) = f(x) + 0 = f(x)$ for all x. Similarly $(0 + f)(x) = f(x)$ for all x in $\mathbb{R}$.

Finally, every function $f \in \mathscr{F}$ has an "inverse" with respect to the function 0, namely, the function $-f$ defined by $(-f)(x) = -[f(x)]$ for all $x \in \mathbb{R}$. (For example, if $f(14) = -3$, then $(-f)(14) = -(-3) = 3$.) In the general case, check that

$f + (-f) = (-f) + f = 0$ for any $f \in \mathscr{F}$

by showing that $(f + (-f))(x) = ((-f) + f)(x) = 0(x) = 0$ for each x in $\mathbb{R}$.

This example marks the end of the current chapter. Our next task will be to extract the common mathematical features of all these seemingly different examples and proceed with the study of group theory.

EXERCISES

1. Investigate the assertions made above and show that $(\mathscr{F}, +)$ satisfies all the properties we have discussed.
2. Investigate the properties discussed for the system $(\mathscr{F}, \cdot)$ where $\cdot$ is the product of functions ($f \cdot g$ is the function defined by $(f \cdot g)(x) = f(x)g(x)$ for any f, g in $\mathscr{F}$).
3. Make a list of the properties common to all the systems we have looked at in this chapter.

2
The definition of a group

In the previous chapter we examined in detail many different "algebraic systems." That is, we have been looking at sets that had an operation—a way of combining any pair of elements in the set to get a third element in the set—associated with them. Thinking back over all the different systems, you may have a sense of how different they all are. You might even be wondering why they are all included in the same chapter.

The reason is that for the purposes of group theory those systems (along with many others) have more in common with each other than they have differences. One of the most beautiful and useful features of pure mathematics is to be able to see and study the similarities among seemingly different systems. To give a nonmathematical example of this type of thinking, look at how we call large numbers of diverse things "fruits" or "vegetables" in order to emphasize certain similarities. In the same way, mathematicians will label various large sets of mathematical systems in order to emphasize particular similarities which are agreed to be beautiful or useful.

There is more to it than that. Many times the common properties of these systems are so powerful and interesting that their study becomes more important than the investigation of the original systems. At the same time, that study gives detailed and useful information about those systems. Group theory is a perfect example of this.

We start by noting the properties common to the systems of the previous chapter, and defining a group.

Let us reflect on those systems. What are they? They are all sets containing objects—numbers, functions, matrices, symbols, and so on. In addition, along with each set we had a rule for combining pairs of elements to get a third element of the set. We make a definition.

DEFINITION
If S is a set, then a function

$$*: S \times S \to S$$

is called a **binary operation** on S.

This function $*$ acts on $S \times S$ as follows:

$*((a, b)) = a * b,$ an element of S.

For example, addition, $+$, is a binary operation on the set $\mathbb{R}$ of real numbers. It is defined by

$+((a, b)) = a + b.$

Specifically

$+((3, 5)) = 3 + 5 = 8$
$+((-1, 4)) = -1 + 4 = 3.$

Conventionally we use the notation $3 + 5 = 8$ rather than

$+((3, 5)) = 8.$

We will also use the notation $a * b$ instead of $*((a, b))$.

Multiplication on $\mathbb{R}$, matrix multiplication on $N_{2 \times 2}$, and function addition on $\mathscr{F}$ are other examples of binary operations.

DEFINITION
An **algebraic system** $(S, *)$ is a set S together with a binary operation $*$ defined on it.

For example, the real numbers $\mathbb{R}$ with the operation of addition is the algebraic system $(\mathbb{R}, +)$; the set of all 2×2 matrices with the operation of matrix multiplication gives the system $(N_{2 \times 2}, \cdot)$.

The examples of the last chapter have similarities other than the fact that they are all algebraic systems. We saw that each system satisfied an "associative law" relative to its binary operation. For the systems with addition as the operation, this law was $(x + y) + z = x + (y + z)$ for all x, y, and z in the set. For the systems with multiplication as the operation,

$(x \cdot y) \cdot z = x \cdot (y \cdot z)$. Generally, a system $(S, *)$ is **associative** if $(x * y) * z = x * (y * z)$ for all x, y, and z in S. Associativity is a similarity or common property of our examples.

Look at the 2×3 matrices under addition, $(M_{2 \times 3}, +)$. The matrix $0_{2 \times 3} = \begin{pmatrix} 0 & 0 & 0 \\ 0 & 0 & 0 \end{pmatrix}$ has the property that $M + 0_{2 \times 3} = 0_{2 \times 3} + M = M$ for any 2×3 matrix, M. If we look at the nonzero real numbers under multiplication, $(\mathbb{R} - \{0\}, \cdot)$ we see that the number 1 satisfies the property $1 \cdot x = x \cdot 1 = x$ for every nonzero real number x.

In a system $(S, *)$ an element $e \in S$ is an **identity element** if for any s in S,

$$s * e = e * s = s.$$

These elements, $0_{2 \times 3}$, 1, are the identity elements in their respective systems. In fact, every system of Chapter 1 has its own identity element. This is another common property of these systems.

The final property that we will use to characterize the notion of a group is the existence of inverses in the set. If $(S, *)$ is an algebraic system *with identity e*, and $s \in S$, then an element t in S is said to be an **inverse for s** (with respect to the identity e) if

$$s * t = t * s = e.$$

Let us look at the examples. Start with the 2×3 matrices under matrix addition. What is the identity element? The matrix $0_{2 \times 3}$ satisfies the necessary property. Now, if M is any 2×3 matrix, then the 2×3 matrix $(-M)$ satisfies $M + (-M) = (-M) + M = 0_{2 \times 3}$. Therefore $(-M)$ is an inverse for M relative to the identity $0_{2 \times 3}$. For example,

$$\begin{pmatrix} 2 & 3 & 7 \\ -2 & 4 & -6 \end{pmatrix} + \begin{pmatrix} -2 & -3 & -7 \\ 2 & -4 & 6 \end{pmatrix} = \begin{pmatrix} 0 & 0 & 0 \\ 0 & 0 & 0 \end{pmatrix}$$

so $\begin{pmatrix} 2 & 3 & 7 \\ -2 & 4 & -6 \end{pmatrix}$ is the inverse of $\begin{pmatrix} -2 & -3 & -7 \\ 2 & -4 & 6 \end{pmatrix}$.

Similarly, if we look at the nonzero real numbers under multiplication, for any x the number $1/x$ satisfies $x \cdot 1/x = 1/x \cdot x = 1$ where 1 is the identity element of this set. Therefore every x in $\mathbb{R} - \{0\}$ has an inverse $1/x$ also in $\mathbb{R} - \{0\}$. As another example, any x in $\{\mathbb{R}, +\}$ has its inverse $(-x)$ since $x + (-x) = (-x) + x = 0$. If you look back to the last chapter, you will see that in each of our examples except Example 1.4, every element of the set had an inverse element that was also in the set.

Now do you have an idea of the structural similarities among all those diverse systems? Our goal in the study of group theory is to make effective use of these similarities.

We start with the definition of a group.

DEFINITION

If G is a set and $*$ is a binary operation on G, then the system $(G, *)$ is a **group** if it satisfies all of the following properties:
1. $x * (y * z) = (x * y) * z$ for all x, y, z in G.
2. There is an element $e \in G$ with the property $e * x = x * e = x$ for all x in G.
3. For every x in G, there is an element x^{-1} in G with the property $x * x^{-1} = x^{-1} * x = e$.

(Note that $*$ being a binary operation insures that $x * y \in G$ for all x and y in G.)

From the last chapter the systems of Examples 1.2, 1.3, 1.5, 1.6, and 1.7 are groups because they satisfy all the properties in the definition.

Example 1.4 is not a group because it is not true that every element of the set has an inverse in the set.

DEFINITION

1. If $(G, *)$ is a group and G has an infinite number of elements, then $(G, *)$ is said to be a group of **infinite order.**
2. If G has a finite number of elements, then the **order** of the group $(G, *)$ is the number of elements in G.
3. We write $|G|$ = order of G.

The groups we studied in Chapter 1 have infinite order. In Chapter 3 we will study some groups with finite orders.

A word here about notation: In the last chapter we sometimes denoted the inverse of an element by something other than x^{-1}. For example, we called $-x$, the *additive* inverse of a real number x, and we called $-M$ the additive inverse of a matrix M. This distinction is necessary due to the fact that some sets, like the real numbers or matrices, have more than one operation defined on them and we must distinguish inverses with respect to the different operations. Therefore we write $-x$ and x^{-1} to denote the additive and multiplicative inverses respectively. For example, the additive inverse of 3 is -3, while the multiplicative inverse of 3 is $3^{-1} = \frac{1}{3}$.

The next definition specifies an important special class of groups.

DEFINITION
A group $(G, *)$ is called an **abelian group** if

$$x * y = y * x \qquad \text{for all } x, y \in G.$$

That is, $(G, *)$ is abelian if and only if it is commutative.

In Chapter 1 the groups of Examples 1.1, 1.2, 1.3, 1.5, and 1.7 are abelian groups. The group $(N_{2 \times 2}, \cdot)$ of Example 1.6 is the only example of a non-abelian group that we have studied so far, but we will see other non-abelian groups in the next chapter.

EXERCISES

In Exercises 1, 2, and 3 below, verify your answer. That is, if you think the system is a group, prove it; if the system is not a group, demonstrate which properties it fails to satisfy and show why.

1. Let V be the set of all vectors (arrows) in the plane. Vector addition, $+$, is defined by placing the vectors tail to head and taking the resultant vector; that is,

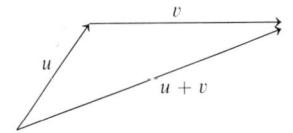

    Is the system $(V, +)$ a group?

2. a. Let $K = \{e, a, b\}$ and let $*$ be an operation on K defined by the following table:

*	e	a	b
e	e	a	b
a	a	e	b
b	b	e	a

    Is $(K, *)$ a group?

    b. Suppose an operation $\circ$ is defined on K by

∘	e	a	b
e	e	a	b
a	a	b	e
b	b	e	a

    Is $(K, \circ)$ a group?

3. Suppose an operation Δ is defined on the integers Z by
$$a \Delta b = 2(a + b).$$
Is (Z, Δ) a group?
4. Prove that there exists *no* real number x whose additive inverse is equal to its multiplicative inverse.
5. Let M be a 2×2 matrix whose additive inverse is equal to its multiplicative inverse. Show that $\det M = 1$, and that M must be of the form
$$\begin{pmatrix} a & b \\ \dfrac{-1-a^2}{b} & -a \end{pmatrix}$$
where a and b are real numbers and $b \neq 0$.

3
Some important groups

Before we go on, it would be a good idea to make a mental list of familiar systems that are groups as well as a list of systems that are not groups. We will look at some additional examples. These will help to clarify the notion as well as to build up your knowledge of certain important groups. We will also call on many of these groups as examples to illustrate the general theory.

The operations on the systems in this chapter are not the usual ones of addition and multiplication. In each example, the set as well as the operation are specifically defined.

EXAMPLE 3.1
This example is more abstract than the preceding ones. Let $H = \{e, a\}$, a set with exactly two elements. Define an operation $\circ$ on H by

$$e \circ e = e, \quad e \circ a = a, \quad a \circ e = a, \quad a \circ a = e.$$

Following our previous pattern, we look at the system $(H, \circ)$, formed by the set H, together with the operation $\circ$. Is it true that for any x and y in H, $x \circ y$ is in H? Verify this by inspecting the definition of the operation.

We can also verify that e acts as an identity:

$$e \circ x = x \circ e = x \quad \text{for any } x \in H$$

(that is for either $x = e$ or $x = a$).

Does each element have an inverse relative to e? To check, $e \circ e = e$ so e is its own inverse, while $a \circ a = e$ so that a is its own inverse.

Finally, does the associative law hold? Is $(x \circ y) \circ z = x \circ (y \circ z)$ for all x, y, z in H? For example, if $x = e, y = a, z = e$, then $(x \circ y) \circ z = (e \circ a) \circ e = a$

Some important groups

(why?), while $x \circ (y \circ z) = e \circ (a \circ e) = a$ also. Convince yourself that $(H, \circ)$ satisfies the associative law; there are seven more cases to be considered.

This system, like the previous ones, satisfies the properties of associativity, existence of an identity, and existence of inverses. Also, the operation or "product" of any two elements of H is again in H. The system $(H, \circ)$ is a group.

EXAMPLE 3.2
Let $G = \{e, a, b, c\}$, a set with 4 elements. Define a product on G by

$\circ$	e	a	b	c
e	e	a	b	c
a	a	e	c	b
b	b	c	e	$a \leftarrow b \circ c = a$
c	c	b	a	e

$c \circ a = b$

By looking at the body of the table, it is easily seen that for any x and y in G, $x \circ y$ is in G. (All the elements in the body of the table represent the products and they are all elements of G.)

Second, we need to check that e is an identity; that is, $e \circ x = x \circ e = x$. This is verified by inspecting the row and the column in the table that represent products by e.

Third, it must be established that every element has an "inverse" with respect to e. Can you see this directly from the table?

Finally, it must also be shown that the system $(G, \circ)$ is associative: $x \circ (y \circ z) = (x \circ y) \circ z$ for any elements $x, y,$ and z in G. We leave this last assertion as an exercise. Try to be creative and shorten the process. It is not necessary to look at all 64 possibilities separately.

This system, $(G, \circ)$ occurs often enough that it has been given a name. It is called the **Klein four-group**.

EXERCISES
1. Show that $(H, \circ)$ is an associative system.
2. Show that the Klein four-group $(G, \circ)$ satisfies the following properties:
   a. If x and y are in G, then $x \circ y \in G$.
   b. $x \circ e = e \circ x = x$ for all x in G.
   c. Every element of G has an <u>inverse</u> in G (relative to e).
   d. $(G, \circ)$ is associative. (Try to avoid looking explicitly at all 64 cases.)
3. Consider the set $K = \{e\}$ with the operation defined by $e \circ e = e$. Which of the properties listed in Exercise 2 does the system $(K, \circ)$ satisfy?

4. Define a product ∘ on the set $S = \{e, a, b\}$ so that the system $(S, \circ)$ satisfies the properties listed in Exercise 2. Verify each of the properties.

Next we will look at some groups arising out of modular arithmetic—the **cyclic** groups. Then we will study groups constructed by rigid motions of a geometric figure and finish with an investigation of groups formed by compositions of a special class of functions.

GROUPS OF INTEGERS MODULO n

EXAMPLE 3.3
Let $G_5 = \{0, 1, 2, 3, 4\}$. Define a binary operation $\oplus$ on G_5 by

$\oplus$	0	1	2	3	4
0	0	1	2	3	4
1	1	2	3	4	0
2	2	3	4	0	1
3	3	4	0	1	2
4	4	0	1	2	3

The operation $\oplus$ is called addition modulo 5 and is denoted by $i \oplus j$ or $i + j \pmod{5}$ for every i and j in G_5.

To show that $(G_5, \oplus)$ is a group we note that the sum of every pair of elements in G_5 is again in G_5. (The fact that the body of the table contains no elements outside of G_5 shows this.)

0 acts as an identity element. This is reflected in the fact that the first row and column of the table mimic the elements; that is, the third element in the first row represents $0 \oplus 2$ and it equals 2; the third element in the first column is $2 \oplus 0 = 2$.

We can also see from the table that each element has an inverse since there is a zero in every row and column. Specifically, 1 and 4 are inverses; 2 and 3 are inverses; 0 is its own inverse.

Associativity of this system is shown in the usual way by showing for every x, y, and z in G_5,

$$(x \oplus y) \oplus z = x \oplus (y \oplus z).$$

We can generalize this group. For any integer $n > 1$, let $G_n = \{0, 1, 2, \ldots, n - 1\}$, and define the binary operation $\oplus$ on G_n by

$$i \oplus j = i + j \pmod{n} = \begin{cases} i + j, & \text{if } i + j < n. \\ i + j - n, & \text{if } i + j \geq n. \end{cases}$$

In G_5, $2 \oplus 1 = 3$, $2 \oplus 2 = 4$, $2 \oplus 3 = 5 - 5 = 0$, $2 \oplus 4 = 6 - 5 = 1$ and we see that we could have defined the operation in G_5 in this way. For another example, $G_2 = \{0, 1\}$ and the operation $\oplus$ is given by $0 \oplus 0 = 0$, $0 \oplus 1 = 1$, $1 \oplus 0 = 1$, $1 \oplus 1 = 2 - 2 = 0$.

To show that $(G_n, \oplus)$ is a group we have to verify all the defining conditions for groups. First, if $i, j \in G_n$, then $i \oplus j \in G_n$ by the way we have defined $\oplus$. Next we need to determine if $(i \oplus j) \oplus k = i \oplus (j \oplus k)$ for all i, j, k in G_n. Well, $i \oplus j = i + j$ or $i + j - n$, so $(i \oplus j) \oplus k$ must be one of $i + j + k$, $i + j + k - n$, or $i + j + k - 2n$ whichever of these is in G_n. Similarly $j \oplus k = j + k$ or $j + k - n$ so $i \oplus (j \oplus k)$ is one of $i + j + k$, $i + j + k - n$, or $i + j + k - 2n$, whichever is in G_n. Since only one of these expressions can be in G_n (why?), $(i \oplus j) \oplus k$ must equal $i \oplus (j \oplus k)$, so G_n is associative.

Third, 0 is the identity element of G_n since $i \oplus 0 = 0 \oplus i = i$ for all $i \in G_n$. If $i \in G_n$, $i \neq 0$, then $n - i$ is the inverse of i in G_n since $i \oplus (n - i) = n - n = 0$; 0 is its own inverse. Since $(G_n, \oplus)$ satisfies all the defining properties, for every positive integer n, $(G_n, \oplus)$ is a group.

There are many applications of these groups. $(G_2, \oplus)$ gives us binary arithmetic and is used in computers. A clock is based on a combination of groups $(G_{12}, \oplus)$ and $(G_{60}, \oplus)$. For another example, consider a wheel with n different positions equally spaced on it—for example, a wheel in a combination lock with 10 positions. If the places are labeled $0, 1, 2, \ldots, n - 1$, the motions of the wheel always take a position into a position i places along, for some i, $0 \leq i < n$. If we have two successive motions, i and j, then their combination gives a motion of $i \oplus j$, and the group G_n describes the possible motions of such a wheel.

EXERCISES

1. Construct a table for the operation $\oplus$ in $G_4 = \{0, 1, 2, 3\}$.
2. For every i in $G_6 = \{0, 1, 2, 3, 4, 5\}$, compute all possible "sums" of i with itself. For example, $2, 2 \oplus 2 = 4, 2 \oplus 2 \oplus 2 = 0, 2 \oplus 2 \oplus 2 \oplus 2 = 2$, and so on. Can you see a pattern in each case?
3. a. Construct a table for the set $\{0, 2, 4\}$ under the operation $\oplus$ defined in G_6. Is this system a group? Prove your answer. What about the set $\{0, 3\}$ under the same operation? The set $\{0, 1, 3\}$?
   b. Make a conjecture concerning this type of result in the group G_n for any positive integer n.
4. Show that in the group $(G_n, \oplus)$ it is true that

$$i \oplus j = j \oplus i$$

for any $i, j \in G_n$.

GROUPS ARISING IN GEOMETRICAL CONTEXTS
EXAMPLE 3.4
Consider a square with vertices A,B,C,D. We will be looking at **symmetries** of this square. A symmetry is a rigid motion of the square which maps vertices to vertices and preserves distances between points. One such symmetry is given by

$$\begin{array}{cc} A & B \\ D & C \end{array} \quad \xrightarrow{r_1} \quad \begin{array}{cc} D & A \\ C & B \end{array}$$

We will refer to this motion (rotation through 90°) as r_1. The other symmetries of the square are

$$\begin{array}{cc} A & B \\ D & C \end{array} \xrightarrow{r_2} \begin{array}{cc} C & D \\ B & A \end{array} \qquad \text{rotation through } 180°;$$

$$\begin{array}{cc} A & B \\ D & C \end{array} \xrightarrow{r_3} \begin{array}{cc} B & C \\ A & D \end{array} \qquad \text{rotation through } 270°;$$

$$\begin{array}{cc} A & B \\ \hline D & C \end{array} \, l \xrightarrow{h} \begin{array}{cc} D & C \\ A & B \end{array} \qquad \text{reflection about the } horizontal \text{ line } l;$$

$$\begin{array}{c|c} A & B \\ D & C \end{array} \xrightarrow{v} \begin{array}{cc} B & A \\ C & D \end{array} \qquad \text{reflection about the } vertical \text{ line } l';$$

(with l' shown on the left square)

$$\begin{array}{cc} A & B \\ D & C \end{array} \xrightarrow{d_1} \begin{array}{cc} C & B \\ D & A \end{array} \qquad \text{reflection about the diagonal } DB;$$

$$\begin{array}{cc} A & B \\ D & C \end{array} \xrightarrow{d_2} \begin{array}{cc} A & D \\ B & C \end{array} \qquad \text{reflection about the diagonal } AC; \text{ and}$$

$$\begin{array}{cc} A & B \\ D & C \end{array} \xrightarrow{I} \begin{array}{cc} A & B \\ D & C \end{array} \qquad \text{no movement (or rotation through } 360°).$$

The motion $\begin{array}{cc} A & B \\ D & C \end{array} \rightarrow \begin{array}{cc} A & B \\ C & D \end{array}$ is not a symmetry since the second square cannot be obtained from the first by a rigid motion. For example, the distance between A and D has been changed.

An operation is defined on the set of all symmetries of the square $\{I, r_1, r_2, r_3, h, v, d_1, d_2\} = D_4$ by considering the resultant when one motion is followed by another.

Some important groups

For example,

```
A ┌───┐ B      D ┌───┐ C      B ┌───┐ C
  │   │   h→     │   │   d₁→    │   │
D └───┘ C      A └───┘ B      A └───┘ D
```

or $h \circ d_1 = r_3$.

The table below contains all products obtained from this operation on the elements of D_4.

∘	I	r_1	r_2	r_3	h	v	d_1	d_2
I	I	r_1	r_2	r_3	h	v	d_1	d_2
r_1	r_1	r_2	r_3	I	d_1	d_2	v	h
r_2	r_2	r_3	I	r_1	v	h	d_2	d_1
r_3	r_3	I	r_1	r_2	d_2	d_1	h	v
h	h	d_2	v	d_1	I	r_2	r_3	r_1
v	v	d_1	h	d_2	r_2	I	r_1	r_3
d_1	d_1	h	d_2	v	r_1	r_3	I	r_2
d_2	d_2	v	d_1	h	r_3	r_1	r_2	I

We can see that the system $(D_4, \circ)$ is a group since:
1. The "product" of any two elements of D_4 is in D_4. (The body of the table consists of elements of D_4.)
2. I acts as an identity element. (The first row and column of the table mimic the elements.)
3. Each element of D_4 has an inverse. (I appears in each row and column of the table. Specifically r_1 and r_3 are inverses; h is its own inverse; so is v; and so on.)
4. Associativity is shown in the usual way—by showing that $(a \circ b) \circ c = a \circ (b \circ c)$ for any three elements a, b, and c chosen from D_4.

The group $(D_4, \circ)$ is called the **dihedral** group of the square. In general, a dihedral group $(D_n, \circ)$ consists of all symmetries of a regular polygon with n sides. The operation $\circ$ is performed by following one operation with another.

Many beautiful and complex groups arise in such geometrical contexts and some of them are very important to chemists and physicists. For example, if we consider a crystal in three-dimensional space and look at all symmetries of that crystal, then these motions, under the operation of following one motion with another, form a group. Depending on the complexity or simplicity of the crystal, the groups can get very complicated, but are quite beautiful as well as important.

Before we continue, let us observe that every *finite* group we have studied before this has been commutative; that is, every one has satisfied

$x \circ y = y \circ x$ for all group elements of x and y. $(D_4, \circ)$ is our first example of a finite noncommutative group. (The group of matrices under multiplication is also noncommutative, but it has an infinite number of elements.) The group S_3 that we will study next is also finite and noncommutative.

EXERCISES

1. Determine the dihedral group D_3 of an equilateral triangle

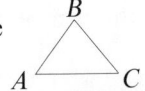

   Exhibit a multiplication table for this group. What is the identity? Which pairs of elements are inverses? Is D_3 commutative?
2. a. Determine the symmetries S of a rectangle

   b. Develop a table for the composition of these symmetries.
   c. Is this system $(S, \circ)$ a group? Justify your answer.
3. Let $\tau_{a,b}$ be the translation which maps the point (x, y) in the plane to the point $(x + a, y + b)$; that is, $\tau_{a,b}(x, y) = (x + a, y + b)$. On $T = \{\tau_{a,b} | a \text{ and } b \text{ are real numbers}\}$, we define an operation $*$ by

$$\tau_{a,b} * \tau_{c,d} = \tau_{a+c, b+d}.$$

   Is $(T, *)$ a group? Justify your answer.
4. a. Let S be the subset of T (from Exercise 3) consisting of those $\tau_{a,b}$ such that a and b are integers. Show that $(S, *)$ is a group.
   b. If $U = \{\tau_{0,b} | b \text{ is a real number}\}$, is $(U, *)$ a group? Why or why not?
   c. $V = \{\tau_{a,b} | a > 0\}$ is $(V, *)$ a group? Explain.

PERMUTATION GROUPS

The permutation groups are groups of functions under the operation of composition. They provide us with important examples that illustrate some of the concepts in group theory. We will illustrate with a particular example.

EXAMPLE 3.5

Let $X = \{1, 2, 3\}$. We consider the set S_3 of all correspondences of X with itself; that is, all bijective (one-to-one and onto) functions $f: X \to X$. One element of S_3 is the function f_1 defined by $f_1(1) = 2$, $f_1(2) = 1$, $f_1(3) = 3$. To make things easier, we agree to denote f_1 by an array signifying what f_1 does to each of the elements of X. We write

$$f_1 = \begin{pmatrix} 1 & 2 & 3 \\ 2 & 1 & 3 \end{pmatrix}.$$

Some important groups

Reading the array vertically, $1 \to 2, 2 \to 1$, and $3 \to 3$. Similarly, if $f_2(1) = 1$, $f_2(2) = 3, f_2(3) = 2$, then we write

$$f_2 = \begin{pmatrix} 1 & 2 & 3 \\ 1 & 3 & 2 \end{pmatrix}.$$

How large is S_3? How many correspondences are there of X with itself? We can write them all out:

$$f_0 = \begin{pmatrix} 1 & 2 & 3 \\ 1 & 2 & 3 \end{pmatrix}, \quad f_1 = \begin{pmatrix} 1 & 2 & 3 \\ 2 & 1 & 3 \end{pmatrix}, \quad f_2 = \begin{pmatrix} 1 & 2 & 3 \\ 1 & 3 & 2 \end{pmatrix}$$

$$f_3 = \begin{pmatrix} 1 & 2 & 3 \\ 2 & 3 & 1 \end{pmatrix}, \quad f_4 = \begin{pmatrix} 1 & 2 & 3 \\ 3 & 1 & 2 \end{pmatrix}, \quad f_5 = \begin{pmatrix} 1 & 2 & 3 \\ 3 & 2 & 1 \end{pmatrix}.$$

There is a way of seeing that there can only be 6 such correspondences. For, in a correspondence, 1 must be mapped to either 1, 2, or 3 so there are exactly 3 possible images for 1. Then, having chosen an image for 1, the symbol 2 can be mapped to either of the two remaining symbols (exactly 2 possible images). Having chosen the image for 1 and 2, then 3 must be mapped to the remaining element of X. Since each of these choices gives a different function, there are $3 \cdot 2 \cdot 1 = 6$ possible correspondences or **permutations** of the elements of X.

The operation that we consider on S_3 is that of function composition. The composition, $f_2 \circ f_3$, is the function defined by

$(f_2 \circ f_3)(1) = f_2(f_3(1)) = f_2(2) = 3$
$(f_2 \circ f_3)(2) = f_2(f_3(2)) = f_2(3) = 2$
$(f_2 \circ f_3)(3) = f_2(f_3(3)) = f_2(1) = 1.$

We see that $f_2 \circ f_3 = f_5$ since these two functions act the same on all the elements of X.

The operation of composition can be accomplished easily with our new notation. We do this as follows: To compose $f_2 \circ f_3$, write them down next to each other and follow the image of each element from right to left. For example,

$$f_2 \circ f_3 = \begin{pmatrix} 1 & 2 & 3 \\ 1 & 3 & 2 \end{pmatrix} \circ \begin{pmatrix} 1 & 2 & 3 \\ 2 & 3 & 1 \end{pmatrix} = \begin{pmatrix} 1 & 2 & 3 \\ 3 & 2 & 1 \end{pmatrix}.$$

Following the arrow, we see f_3 takes $1 \to 2$, f_2 takes $2 \to 3$, so $f_2 \circ f_3$ takes $1 \to 3$. Similarly f_3 takes $2 \to 3$, f_2 takes $3 \to 2$, so $f_2 \circ f_3$ takes $2 \to 2$. Finally f_3 takes $3 \to 1$, f_2 takes $1 \to 1$, so $f_2 \circ f_3$ takes $3 \to 1$. As another example,

$$f_3 \circ f_5 = \begin{pmatrix} 1 & 2 & 3 \\ 2 & 3 & 1 \end{pmatrix} \circ \begin{pmatrix} 1 & 2 & 3 \\ 3 & 2 & 1 \end{pmatrix} = \begin{pmatrix} 1 & 2 & 3 \\ 1 & 3 & 2 \end{pmatrix} = f_2.$$

With a little practice these computations become very fast and easy.

The table for S_3 under the operation of composition is as follows:

$\circ$	f_0	f_1	f_2	f_3	f_4	f_5
f_0	f_0	f_1	f_2	f_3	f_4	f_5
f_1	f_1	f_0	f_3	f_2	f_5	f_4
f_2	f_2	f_4	f_0	f_5	f_1	f_3
f_3	f_3	f_5	f_1	f_4	f_0	f_2
f_4	f_4	f_2	f_5	f_0	f_3	f_1
f_5	f_5	f_3	f_4	f_1	f_2	f_0

A moment's glance shows that f_0 is an identity element for S_3 relative to this composition and that every element has an inverse relative to f_0. It is not too difficult to show that $(S_3, \circ)$ is an associative system. We leave for the exercises the fact that S_3 is indeed a group as well as some of the interesting properties of this group.

To generalize this type of group, we fix a positive integer n and consider a set $X = \{1, 2, \ldots, n\}$ along with the set S_n of all one-to-one correspondences $f: X \to X$. As in S_3, each function f gives a permutation (rearrangement), $(j_1, j_2, \ldots, j_n)$ of the elements, $(1, 2, \ldots, n)$, of X. (That is, $f(i) = j_i$.) We write

$$f = \begin{pmatrix} 1 & 2 & 3 & \cdots & n \\ j_1 & j_2 & j_3 & \cdots & j_n \end{pmatrix}.$$

The operation on the set S_n is composition $\circ$ and is accomplished the same way as in S_3. $(S_n, \circ)$ can be shown to be a group for any positive integer n.

At the end of this part of the book we will return to the groups S_n and give a more complete exposition of them. In the meantime, they (S_3 in particular) are important examples of finite groups which are noncommutative.

EXERCISES

1. Calculate the compositions

a. $\begin{pmatrix} 1 & 2 & 3 \\ 3 & 1 & 2 \end{pmatrix} \circ \begin{pmatrix} 1 & 2 & 3 \\ 1 & 3 & 2 \end{pmatrix}$ b. $\begin{pmatrix} 1 & 2 & 3 \\ 1 & 3 & 2 \end{pmatrix} \circ \begin{pmatrix} 1 & 2 & 3 \\ 3 & 1 & 2 \end{pmatrix}$

c. $\left[\begin{pmatrix} 1 & 2 & 3 \\ 3 & 1 & 2 \end{pmatrix} \circ \begin{pmatrix} 1 & 2 & 3 \\ 3 & 2 & 1 \end{pmatrix}\right] \circ \begin{pmatrix} 1 & 2 & 3 \\ 2 & 1 & 3 \end{pmatrix}$

in S_3.

2. Show that S_3 is a group.
3. In S_3 calculate the compositions, $f_3^2 = f_3 \circ f_3, f_3^3, f_3^4,$ and so on. Make a table for all the powers of f_3 under composition. Is the set $\{f_3^n | n$ is a positive integer$\}$ a group under the operation of composition?
4. Define S_2 and show it is a group.
5. How large is S_4? Show S_4 is a group without looking at all the elements specifically.
6. In the group $(S_n, \circ)$, what is the identity element? If

$$f = \begin{pmatrix} 1 & 2 & \cdots & n \\ j_1 & j_2 & \cdots & j_n \end{pmatrix}$$

find an expression for f^{-1}.

4
Development of elementary group properties

In this chapter we begin to develop the subject of elementary group theory. We shall prove many facts (theorems) about abstract groups which follow from parts 1, 2, and 3 of the definition of a group in Chapter 2. Most of this will be done without reference to specific examples. This exercise is fun and can be very exciting. The result is detailed information that can be applied to any specific example of a group that may arise.

Our first step will be to prove a number of elementary results. As we progress, we will give many examples illustrating the concepts, but the theorems and proofs hold for groups in general, not only for the examples.

We will be looking at results dealing with the properties of identities, inverses, and associativity.

Before continuing we need to state formally that, for a group $(G, \circ)$ and elements x, y in G, the symbol

$$x = y$$

means that x and y are (possibly different) symbols for the same element of the set. We will also be using some basic logical principles such as
1. If $x = y$ and $y = z$, then $x = z$.
2. If $x = y$, then $a \circ x = a \circ y$ and $x \circ a = y \circ a$ for all $a, x, y,$ and z in G.

Our first result is a very useful one; it will help us in proving many further results. The proof involves all three axioms—associativity, existence of an identity, and existence of inverses—in the definition of a group.

THEOREM 4.1: CANCELLATION LAWS
In a group $(G, \circ)$,
1. If $a \circ x = a \circ y$, then $x = y$. (left cancellation law)
2. If $x \circ a = y \circ a$, then $x = y$. (right cancellation law)

Development of elementary group properties

Proof of 1: For elements a, x, and y of G, suppose that

$$a \circ x = a \circ y.$$

The element a has an inverse, a^{-1}; therefore

$$a^{-1} \circ (a \circ x) = a^{-1} \circ (a \circ y).$$

Reassociating, we get

$$(a^{-1} \circ a) \circ x = (a^{-1} \circ a) \circ y$$
$$e \circ x = e \circ y$$

since $a \circ a^{-1} = e$, and using the property of the identity,

$$x = y. \blacksquare$$

The proof of 2 is similar and is left as an exercise.

The next results give us more information about the identity element. The cancellation laws are used in the proofs. First, we show that any given group can have only one identity element.

THEOREM 4.2
The identity element of a group is unique.

Proof: Suppose e and f are both identity elements for the group $(G, \circ)$; then for any x in G,

$$x \circ e = e \circ x = x$$

and

$$x \circ f = f \circ x = x.$$

Therefore

$$x \circ e = x \circ f$$

and by the cancellation laws

$$e = f.$$

It follows that e and f are merely different names for the same element and that any element in G that satisfies the property of an identity must be equal to e. $\blacksquare$

THEOREM 4.3
If $(G, \circ)$ is a group and x is an element of G satisfying $x \circ x = x$, then x is the identity element of the group.

Proof: Suppose $(G, \circ)$ is a group and x is an element of G with the property

$$x \circ x = x.$$

Since $(G, \circ)$ is a group, there is an identity $e \in G$. Therefore

$$x \circ e = x$$

and we have

$$x \circ x = x \circ e.$$

By the first cancellation law,

$$x = e. \quad \blacksquare$$

Note: In a system $(S, \circ)$, if an element $x \in S$ has the property $x \circ x = x$, then x is called **idempotent.** The previous theorem says that, in a group, the only idempotent element is the identity.

Let us investigate what these results mean in terms of the groups with which we are familiar.

The cancellation laws are a very powerful tool for working with groups. We have already seen evidence of that in the proofs of Theorems 4.2 and 4.3.

Theorem 4.2 guarantees that each group has exactly one identity. For example, in $(\mathbb{Z}, +)$, 0 is the only identity. The real number 1 is the unique identity in $(\mathbb{R} - \{0\}, \cdot)$. The matrix $I = \begin{pmatrix} 1 & 0 \\ 0 & 1 \end{pmatrix}$ is the *only* one in $(N_{2 \times 2}, \cdot)$ with the property that $AI = IA = A$ for every $A \in N_{2 \times 2}$.

Let us look at a special case of Theorem 4.3 in the reals. Let x be a real number with the property

$$x^2 = x.$$

Solving this, we get *two* solutions, $x = 0$ and $x = 1$. Does this contradict the theorem? No, because under multiplication $(\mathbb{R}, \cdot)$ is not a group, but $(\mathbb{R} - \{0\}, \cdot)$ is. That is, the number 0 is *not* an element of this group. The fact that 0 as well as 1 satisfies this equation does not contradict the theorem. As a matter of fact, our calculations have shown that the identity 1 is the only element of $(\mathbb{R} - \{0\}, \cdot)$ that satisfies this equation. This illustrates, rather than contradicts, the meaning of the theorem.

We know that in a group, each element has an inverse. Our next theorem asserts that the inverse of an element is unique.

Development of elementary group properties

THEOREM 4.4

Every element in a group has exactly one inverse.

Proof: Let $(G, \circ)$ be a group and let $x \in G$. Suppose x has two inverses, y and z. Then

$$x \circ y = y \circ x = e \quad \text{since } y \text{ is an inverse}$$

and

$$x \circ z = z \circ x = e \quad \text{since } z \text{ is an inverse.}$$

Since both equations have the same value e,

$$x \circ y = x \circ z.$$

Again, using the cancellation laws,

$$y = z.$$

We conclude that x must have only one inverse. ∎

Theorems 4.2 and 4.4 justify our notation. First, since there is only *one* identity element in any group, it makes sense to use a single symbol, e, to denote the element. Second, since every element has only one inverse, it is not at all ambiguous to use some symbol related to x, like x^{-1}, to denote the unique inverse of x. Some confusion might arise when we are dealing with a known group, like the integers under addition whose identity element already has a common name, 0, and where $-x$ usually denotes the inverse of x. However, since these groups are familiar, the meaning should be clear in context.

These theorems illustrate a good point—that one often has to work to prove what seems obvious. After all, in any familiar groups we can think of, there is only one identity element and inverses are unique. Why fuss so over abstract groups? The answer is simply that we cannot assume anything to be true that is not either an axiom or provable from the axioms. That is the nature of abstract mathematics and it is that which makes mathematical statements so certain. Everything has always been checked and there can be no surprises in store for us as long as our logical inferences are done correctly. That is, it is not conceivable that the development of a new microscope or telescope or any other technological advance will force us to revise our group-theoretic results. These are, after all, logical deductions from axioms and, as such, are inimitable.

By now you should be getting some feel for working with arbitrary elements in a group by manipulating algebraic symbols that represent the elements. The *rules* of manipulation are our axioms, plus, of course, any theorems we have already proved.

It would be useful at this point if you turned back to some examples of groups and integrated Theorems 4.1 through 4.4 in those groups. What are the unique identities in each group? How is the inverse of each element obtained? What do the cancellation laws mean in these systems? In general, it is a good idea to try to understand a result in a specific, familiar system before hoping to fully absorb the theorem in its most abstract setting.

We continue with an important result based on the axiom of associativity. The proof is done by mathematical induction. The idea behind mathematical induction is that a statement can be proved true for every positive integer by first showing that the statement is true for the integer 1 and then showing how to get the result for a positive integer given that it is true for the previous integer (that is, true for $k \Rightarrow$ true for $k + 1$). Watch for this pattern as we do the proof. If the statement is true for 1, then it is true for $2 = 1 + 1$; it is true for $3 = 2 + 1$, for $4 = 3 + 1$, and so on. Intuitively, the statement is true for any positive integer.

THEOREM 4.5: THE GENERALIZED ASSOCIATIVE LAW

If $(G, \circ)$ is a group, then, for every positive integer n and elements $g_1, g_2, \ldots, g_n$ in G, the product $g_1 \circ g_2 \circ \cdots \circ g_n$ is uniquely defined.

Note: Since we have specified the order of the elements, the only difference between any two products would be the way in which they are associated (that is, the placement of the parentheses). This theorem actually states that the associative law can be extended to cover products of any number of group elements.

Proof: We proceed by induction. The theorem is trivially true for $n = 1$: $g_1 = g_1$. (It also holds immediately for $n = 2$: $g_1 \circ g_2 = g_1 \circ g_2$. The case $n = 3$ is true by the associative law.)

Assume the theorem is true for all positive integers n up to and including $n = k$, where k is some fixed positive integer. That is, for $r \leq k$, any two products of any group elements $g_1, g_2, \ldots, g_r$ in the same order have the same value. Now choose $k + 1$ elements from G, $g_1, g_2, \ldots, g_{k+1}$ and consider two products of these elements with different placement of the parentheses. Call one product a and the other b. (For example, a could equal $(g_1 \circ g_2) \circ (g_3 \circ \cdots \circ g_{k+1})$ and b might be $g_1 \circ (g_2 \circ \cdots \circ g_{k+1})$. Note that the second factor in both products is unique by the induction hypothesis.)

Development of elementary group properties

Let $a = f_1 \circ f_2$ where f_1 is a product of elements $g_1, g_2, \ldots, g_r$ and f_2 is a product of $g_{r+1}, g_{r+2}, \ldots, g_{k+1}$ for some r, $1 \le r \le k$. Since, by the induction hypothesis, any two products of $g_1, \ldots, g_r$ in that order are equal, we can let $f_1 = g_1 \circ (g_2 \circ \cdots \circ g_r)$. Using this association,

$$\begin{aligned} a &= f_1 \circ f_2 \\ &= [g_1 \circ (g_2 \circ \cdots \circ g_r)] \circ f_2 \\ &= g_1 \circ [(g_2 \circ \cdots \circ g_r) \circ f_2] \quad \text{(by the associative law)} \\ &= g_1 \circ [g_2 \circ \cdots \circ g_{k+1}] \end{aligned}$$

where $g_2 \circ \cdots \circ g_{k+1}$ is unique by our hypothesis since there are only k elements in this product.

In the same way, we let $b = h_1 \circ h_2$ where $h_1 = g_1 \circ \cdots \circ g_s$ and $h_2 = g_{s+1} \circ \cdots \circ g_{k+1}$ for some s, $1 \le s \le k$, $s \ne r$. Reasoning as we did before, we can write $h_1 = g_1 \circ (g_2 \circ \cdots \circ g_s)$ and therefore

$$\begin{aligned} b = h_1 \circ h_2 &= [g_1 \circ (g_2 \circ \cdots \circ g_s)] \circ h_2 \\ &= g_1 \circ [(g_2 \circ \cdots \circ g_s) \circ h_2] \\ &= g_1 \circ [g_2 \circ \cdots \circ g_{k+1}] \end{aligned}$$

and therefore $a = b$.

Invoking the induction axiom, the product $g_1 \circ g_2 \circ \cdots \circ g_n$ is uniquely defined for all positive integers n and for all elements $g_1, g_2, \ldots, g_n$ chosen from a group G. ∎

COROLLARY

If $(G, \circ)$ is a group and $g \in G$, then g^n is well defined.

Proof: Exercise 8.

Now that we know that $g_1 \circ g_2 \circ \cdots \circ g_n$ is well defined, the next theorem tells us how to find its inverse.

THEOREM 4.6

Let $(G, \circ)$ be a group and, for any positive integer n, let $g_1, g_2, \ldots, g_n \in G$; then

$$(g_1 \circ g_2 \circ \cdots \circ g_n)^{-1} = g_n^{-1} \circ g_{n-1}^{-1} \circ \cdots \circ g_1^{-1}.$$

Proof: Again we proceed by induction. For $n = 1$, $g_1^{-1} = g_1^{-1}$ and the result is trivially true.

Although it is not necessary for the proof, we present the case $n = 2$ to aid our intuition. The theorem states that $(g_1 \circ g_2)^{-1} = g_2^{-1} \circ g_1^{-1}$. To prove this, we must show that the product of $g_1 \circ g_2$ and $g_2^{-1} \circ g_1^{-1}$ is the identity.

$(g_1 \circ g_2) \circ (g_2^{-1} \circ g_1^{-1})$
$\quad = [(g_1 \circ g_2) \circ g_2^{-1}] \circ g_1^{-1}$ (by the associative law)
$\quad = [g_1 \circ (g_2 \circ g_2^{-1})] \circ g_1^{-1}$ (again by the associative law)
$\quad = (g_1 \circ e) \circ g_1^{-1}$ (since $g_2 \circ g_2^{-1} = e$)
$\quad = g_1 \circ g_1^{-1}$ (by property of the identity)
$\quad = e.$

Therefore $(g_1 \circ g_2)^{-1} = g_2^{-1} \circ g_1^{-1}$.

Having shown that the theorem holds for $n = 1$ and 2, we assume that for every k elements $g_1, g_2, \ldots, g_k$ chosen from G,

$$(g_1 \circ g_2 \circ \cdots \circ g_k)^{-1} = g_k^{-1} \circ g_{k-1}^{-1} \circ \cdots \circ g_1^{-1}.$$

For elements $g_1, \ldots, g_{k+1}$ in G, we will show that $(g_1 \circ g_2 \circ \cdots \circ g_{k+1})^{-1} = g_{k+1}^{-1} \circ g_k^{-1} \circ \cdots \circ g_1^{-1}$.

$(g_1 \circ g_2 \circ \cdots \circ g_{k+1}) \circ (g_{k+1}^{-1} \circ g_k^{-1} \circ \cdots \circ g_1^{-1})$
$\quad = [(g_1 \circ g_2 \circ \cdots \circ g_{k+1}) \circ g_{k+1}^{-1}] \circ (g_k^{-1} \circ \cdots \circ g_1^{-1})$
$\quad\quad\quad$ (by the associative law)
$\quad = [(g_1 \circ g_2 \circ \cdots \circ g_k) \circ (g_{k+1} \circ g_{k+1}^{-1})] \circ (g_k^{-1} \circ \cdots \circ g_1^{-1})$
$\quad\quad\quad$ (again by the associative law)
$\quad = [(g_1 \circ g_2 \circ \cdots \circ g_k) \circ e] \circ (g_k^{-1} \circ \cdots \circ g_1^{-1})$
$\quad\quad\quad$ (since $g_{k+1} \circ g_{k+1}^{-1} = e$)
$\quad = (g_1 \circ g_2 \circ \cdots \circ g_k) \circ (g_k^{-1} \circ g_{k-1}^{-1} \circ \cdots \circ g_1^{-1})$
$\quad\quad\quad$ (by property of the identity)
$\quad = e$ (by the induction hypothesis). ∎

COROLLARY

If $(G, \circ)$ is a group and $g \in G$, then

$$(g^n)^{-1} = (g^{-1})^n.$$

Proof: Exercise 9.

Notation: We define $g^{-n} = (g^n)^{-1}$.

In Theorem 4.6 note the reversal of the order of the elements. This is very important in noncommutative groups.

EXAMPLE 4.1

In $(S_3, \circ)$ (Example 3.5, Chapter 3), $f_1 \circ f_4 = f_5$, $f_1^{-1} = f_1$, and $f_4^{-1} = f_3$. Our theorem tells us that

$$(f_1 \circ f_4)^{-1} = f_4^{-1} \circ f_1^{-1} = f_3 \circ f_1 = f_5.$$

Checking with the table, we see that f_5 is the desired inverse. Note that the reverse product $f_1 \circ f_3 = f_2$ which is not f_5^{-1}.

We conclude this chapter by stating an extremely important result—that operations with integral exponents behave the same in any group as they do in the real numbers.

THEOREM 4.7: LAW OF EXPONENTS

If $(G, \circ)$ is a group, $g \in G$, and a and b are integers, then
1. $g^a \circ g^b = g^{a+b}$
2. $(g^a)^b = g^{ab}$.

Proof: Exercises 12 and 13.

EXERCISES

In the following, $(G, \circ)$ will be a group.
1. If $x \in G$, prove that $(x^{-1})^{-1} = x$.
2. If x and $y \in G$ with $x \circ y = e$, the identity in G, prove that $x^{-1} = y$ and $y^{-1} = x$.
3. Prove part 2 of the cancellation laws.
4. Prove or disprove: If $a, b \in G$, then $(a \circ b)^n = a^n \circ b^n$ for every positive integer n.
5. Prove or disprove: If $a, b,$ and $c \in G$ and $a \circ b = b \circ c$, then $a = c$.
6. Let S be a *finite* set with an operation $\circ$ defined on it. If $(S, \circ)$ is associative and both cancellation laws hold, prove $(S, \circ)$ is a group. Show this need not be true if S is infinite.
7. What matrices X in $M_{2 \times 2}$ satisfy $X^2 = X$? How does this relate to Theorem 4.3?
8. Prove the corollary to Theorem 4.5. You may use the theorem.
9. Prove the corollary to Theorem 4.6.
10. If we look at the group $(\mathbb{R}, +)$, the real numbers under addition, what should be our notation for
$$\underbrace{x + x + \cdots + x?}_{n \text{ terms}}$$
11. Use Theorem 4.6 to find the inverse of
$$\begin{pmatrix} 2 & 3 \\ 1 & 2 \end{pmatrix} \cdot \begin{pmatrix} 4 & 1 \\ 7 & 2 \end{pmatrix}$$
in $(N_{2 \times 2}, \cdot)$ (Example 1.6, Chapter 1). Check your result.

12. a. If $a > 0$ and $b > 0$, prove that $g^a \circ g^b = g^{a+b}$.
    b. Prove $g^a \circ g^b = g^{a+b}$ for $a > 0$, $b < 0$.
    c. Do part b for $a < 0$, $b > 0$.
    d. Do part b for $a < 0$ and $b < 0$.

Note: After completing this problem, the first law of exponents has been proved.

13. If $g \in G$, prove that $(g^a)^b = g^{ab}$ for any integers a and b (thus completing the proof of the law of exponents).

Subgroups and cyclic subgroups

SUBGROUPS

We turn now to a very important notion in the study of groups, the idea of a subgroup. This concept, a subsystem of a system, is central to all higher algebraic studies because studying the subsystems can provide valuable information about the system itself.

DEFINITION

If $(G, \circ)$ is a group, and if $H \subseteq G$, then $(H, \circ)$ is called a **subgroup** of $(G, \circ)$ if $(H, \circ)$ is a group.

Notation: If $(H, \circ)$ is a subgroup of $(G, \circ)$, we write $(H, \circ) < (G, \circ)$ or, noting that the operations are the same in both groups, $H < G$. In fact, from this point on, we will usually refer to the group $(G, \circ)$ by the single symbol G, and not refer to the operation explicitly.

Basically, a subgroup of a group consists of a subset which forms a group with the *same operation*.

EXAMPLE 5.1

The system of even integers (those divisible by 2) under addition is a subgroup of the integers under addition. We verify this as follows: Let $E = \{2n \mid n \text{ is an integer}\}$. $E \subseteq Z$, the set of integers. We need to show that E is a group under addition.

1. If $2m$ and $2n$ are in E, then

$$2m + 2n = 2(m + n) \in E$$

since this sum is again even. Therefore we say E is **closed** under $+$.

2. If $2l, 2m, 2n \in E$, then

$$(2l + 2m) + 2n = 2l + (2m + 2n)$$

because $2l, 2m,$ and $2n$ are integers and addition of integers is associative. Note that E is associative because $E \subseteq Z$ and Z is associative. (Why is this?)

3. $0 = 2 \cdot 0 \in E$ so that the system $(E, +)$ has an identity.
4. For $2n \in E$, the additive inverse $-2n \in E$ since it is 2 times the integer $(-n)$.

Since E satisfies the definition of a group and since $E \subseteq Z$, E is a subgroup of Z or $E < Z$.

EXAMPLE 5.2

The set of odd integers $\mathcal{O}$ under addition is not a subgroup of $(Z, +)$ since if $2n + 1$ and $2m + 1 \in \mathcal{O}$, then

$$(2n + 1) + (2m + 1) = 2n + 2m + 2$$

which is an even integer and therefore does not belong to $\mathcal{O}$. $\mathcal{O}$ is not closed under the operation $+$ and the set of odd integers is not a group under the operation $+$. (Note that it would not help to take the odd integers and define an operation on them other than addition in order to form a group. Any group formed that way would not be a subgroup of the integers under addition because the operations would differ.)

EXAMPLE 5.3

Another example of a subgroup of $(Z, +)$ is the set $H_7 = \{7n \mid n \in Z\}$ under addition. The proof of this is similar to the proof in Example 5.1 and will follow from Exercise 1.

EXAMPLE 5.4

Consider the cyclic group G_6 (like those in Example 3.3). The set $G_6 = \{0, 1, 2, 3, 4, 5\}$ has many subsets, but let us consider the subset $H = \{0, 2, 4\}$. Inspecting a table for H under the operation $\oplus$,

$\oplus$	0	2	4
0	0	2	4
2	2	4	0
4	4	0	2

it can be seen that $(H, \oplus)$ is a group.
1. The body of the table consists of nothing but 0, 2, and 4. Therefore we may conclude that H is closed under $\oplus$.
2. Since G_6 is a group, the operation $\oplus$ is associative on any elements of G_6. Since elements of H are also elements of G_6, the operation $\oplus$ on the set H is associative.

3. $0 \in H$ and acts as an identity for H.
4. The elements 4 and 2 are inverses; 0 is its own inverse.

H is a subgroup of G_6, $H < G_6$, since $H \subseteq G$ and H is a group in its own right.

EXERCISES

set of integers.

1. Let H_k be the set of all integers divisible by k; that is, $H_k = \{kn \mid n \in Z\}$.
   a. Prove that H_k is a subgroup of Z.
   b. Show that $H_{2k} < H_k$.
2. a. Let K be the set of all 2×3 matrices with a 0 in the first row, first column. Show that K is a subgroup of $M_{2 \times 3}$.
   b. Name three other subgroups of $M_{2 \times 3}$.
   c. Show that the set of all 2×3 matrices with a -4 in the first row, first column is not a subgroup of $M_{2 \times 3}$.
3. Examine *all* subsets of the Klein four-group and determine which of them are subgroups.
4. a. Which subsets of G_6, containing the element 0, form a subgroup under the operation $\oplus$?
   b. Determine all subgroups of G_6 that contain the element 1.
5. Let G be a group and $H < G$.
   a. Show that the identity in H is the same as the identity in G.
   b. Show that the inverse of any element in H is the same as its inverse in G.
6. Which of the following subsets of $M_{2 \times 2}$ with the operation of addition form subgroups?

   a. $\left\{ \begin{pmatrix} a & b \\ b & a \end{pmatrix} \middle| a \text{ and } b \text{ are real numbers} \right\}$.

   b. $\{A \mid A^{-1} \text{ exists}\}$.

   c. $\{A \mid \text{all elements of } A \text{ are either all positive, all negative, or all zero}\}$.
7. Let G be a group.
   a. Is the empty set a subgroup of G?
   b. Can a subgroup of G have just 1 element?

After reading the examples and doing a few of the exercises, it should become apparent that many of the arguments are similar in every case. For example, the associativity of the operation on any group guarantees the associativity of the operation on any subset of that group. We will prove three theorems which capitalize on these similarities and make it easier to prove that a particular subset of a group is a subgroup.

THEOREM 5.1

If G is a group and H is a nonempty subset of G such that
1. whenever x and y are in H, $x \circ y$ is also in H;
2. whenever x is in H, x^{-1} is also in H; then H is a subgroup of G.

Proof: To prove that H is a subgroup of G, we have to show that the system $(H, \circ)$ satisfies all the properties of the definition of a group. Our assumptions state that (1) $\circ$ is a binary operation on H, and (2) for every element of H, its inverse (in G) is also an element of H.

We need to show that the operation $\circ$ is associative on H and that there is an element in H that acts as an identity.

Since the operation $\circ$ is associative on all of G (because G is a group), it is certainly associative on H, a subset of G. (Why?)

To prove that the identity is in H, we need all of the hypotheses. Since H is nonempty, choose an element $x \in H$. By (2), x^{-1} is also in H and, by (1), $x \circ x^{-1}$ is in H. The identity $e \in H$ since $x \circ x^{-1} = e$.

Therefore, since it satisfies all properties stated in the definition, H is a group and also a subgroup of G. ∎

EXAMPLE 5.5

Consider the group D_4 (Example 3.4) and the set $H = \{I, r_1, r_2, r_3\} \subseteq D_4$. Using Theorem 5.1 we will show that H is a subgroup of D_4.

First, for any two elements in H, their "product" is also in H. This can be verified directly by computing all 16 products or by referring to the group table given on p. 23.

Also, $r_1^{-1} = r_3$, $r_2^{-1} = r_2$, $I^{-1} = I$ so that the inverse of any element in H is also in H.

Theorem 5.1 allows us to state that H is a subgroup of D_4 without checking the other two properties.

THEOREM 5.2

If G is a group and H is a nonempty subset of G with the property that whenever x and y are in H, the element $x \circ y^{-1}$ is also in H, then H is a subgroup of G.

Proof: We will consider a nonempty subset $H \subseteq G$ satisfying the hypothesis of Theorem 5.2 and show that H also satisfies the hypotheses of Theorem 5.1. This will guarantee that $H < G$.

Let x be any element of H. Applying the hypothesis of Theorem 5.2 and taking $y = x$, we have $x \circ x^{-1} = e \in H$.

Again choose any $x \in H$. Since $e \in H$, we apply the hypothesis again and get $e \circ x^{-1} = x^{-1} \in H$. Therefore, for any $x \in H$, x^{-1} is also in H, or H satisfies the second hypothesis of Theorem 5.1.

Next, let x and y be in H. Then $y^{-1} \in H$ (why?) and, by assumption, $x \circ (y^{-1})^{-1} \in H$. But $(y^{-1})^{-1} = y$ so that $x \circ y \in H$ for every x and y in H. Thus H satisfies the first hypothesis of Theorem 5.2 also and therefore $H < G$. ∎

EXAMPLE 5.6

Consider the group $M_{2\times 2} = \left\{ \begin{pmatrix} a & b \\ c & d \end{pmatrix} \middle| a, b, c, d \text{ are real numbers} \right\}$ under the operation of addition. A 2×2 matrix $\begin{pmatrix} a & b \\ c & d \end{pmatrix}$ is called *symmetric* if $b = c$. That is, a 2×2 symmetric matrix is of the form $\begin{pmatrix} a & b \\ b & d \end{pmatrix}$. Let $H = $ the set of all symmetric matrices $= \left\{ \begin{pmatrix} a & b \\ b & d \end{pmatrix} \middle| a, b, d \text{ are real} \right\}$. We want to show that H is a subgroup of $M_{2\times 2}$ with the help of Theorem 5.2.

Consider two elements of H, $x = \begin{pmatrix} a & b \\ b & d \end{pmatrix}$ and $y = \begin{pmatrix} e & f \\ f & g \end{pmatrix}$. The inverse of $\begin{pmatrix} e & f \\ f & g \end{pmatrix}$ in this group is $\begin{pmatrix} -e & -f \\ -f & -g \end{pmatrix}$. The "product" $x \circ y^{-1}$ in this group becomes $\begin{pmatrix} a & b \\ b & d \end{pmatrix} + \begin{pmatrix} -e & -f \\ -f & -g \end{pmatrix}$ which equals $\begin{pmatrix} a-e & b-f \\ b-f & d-g \end{pmatrix}$ which is symmetric and therefore is an element of H. By Theorem 5.2, H is a subgroup of $M_{2\times 2}$.

The last two theorems provide shortcuts that are useful in proving that a subset is a subgroup. They hold for all subsets, whether finite or infinite. For finite subsets, a single condition is sufficient.

THEOREM 5.3

Let G be a group. If H is a finite nonempty subset of G such that whenever x and y are in H, $x \circ y$ is also in H, then $H < G$.

Proof: Since H is assumed finite, we can assume $H = \{x_1, x_2, \ldots, x_n\}$. Let us look at the n products $x_1 \circ x_1, x_1 \circ x_2, x_1 \circ x_3, \ldots, x_1 \circ x_n$. By hypothesis they are all in H. We claim that they are all distinct, for if $x_1 \circ x_i = x_1 \circ x_j$, then the cancellation laws guarantee that $x_i = x_j$ which is a contradiction. Since there are n distinct products in H and n elements in H, each element of H must be one of the products. In particular, x_1 itself must be one of the products, $x_1 \circ x_i = x_1$ for some i. But $x_1 = x_1 \circ e$, and again by the cancellation laws, we conclude that $x_i = e$ and therefore $e \in H$.

Next, for any x_r in H, look at the n products $x_r \circ x_1, x_r \circ x_2, \ldots, x_r \circ x_n$. Again, these are distinct elements of H, so these n elements must be all of H. In particular, one of the products, say $x_r \circ x_s$, must be the element $e \in H$. Thus, we have shown that $x_s = x_r^{-1}$ (why?), and therefore, for every x_r in H, x_r^{-1} is also in H.

The hypothesis of this theorem is the first hypothesis of Theorem 5.1; we have shown that H also satisfies the second hypothesis of Theorem 5.1 and therefore H is a subgroup of G. ∎

The proof of this theorem uses a counting trick that requires that H be finite. If H is not finite, the condition that H is closed under the operation is *not enough* to ensure that H is a subgroup. Exercise 3 asks you for an example to illustrate this.

EXAMPLE 5.7

In the symmetric group on three elements S_3 (Example 3.5) consider the subset $H = \{f_0, f_3, f_4\}$. Using Theorem 5.3 to show that H is a subgroup of S_3, we only need to look at a table for H under the operation $\circ$:

$\circ$	f_0	f_3	f_4
f_0	f_0	f_3	f_4
f_3	f_3	f_4	f_0
f_4	f_4	f_0	f_3

and notice that all the elements in the body of the table are in the set $\{f_0, f_3, f_4\}$. Therefore H satisfies the hypothesis of Theorem 5.3 and is a subgroup of S_3.

EXERCISES

1. Use Theorem 5.1 to determine which of the following subsets of $\mathbb{R} - \{0\}$ are subgroups of $\mathbb{R} - \{0\}$ under the operation of multiplication.
   a. $\{x \mid x \text{ is an integer}\}$ b. $\{x \mid x \text{ is rational}\}$
   c. $\{1\}$ d. $\{x \mid x \text{ is irrational}\}$
2. Use Theorem 5.2 to determine which of the following subsets of D_4 are subgroups.
   a. $\{I, d_1\}$ b. $\{I, d_1, d_2\}$
   c. $\{I, d_1, d_2, h, v\}$ d. $\{I, r_2\}$
3. a. Go through the proof of Theorem 5.3 very carefully and find those points where finiteness of H is necessary; that is, where the proof does not apply to infinite subsets.
   b. Give an example to show that this theorem does not apply to infinite subsets of a group. That is, find a group G, an infinite subset $H \subseteq G$ where $x \circ y \in H$ for every $x, y \in H$, but where H is *not* a subgroup of G.
4. Use Theorem 5.3 to find out which subsets of the cyclic groups G_3, G_4, and G_6 are subgroups of those groups.
5. Show that $\{e\}$ and G are always subgroups of any group G. Subgroups other than these are called **proper** subgroups.

Subgroups and cyclic subgroups 43

6. a. Show that $\left\{\begin{pmatrix} 1 & x \\ 0 & 1 \end{pmatrix} \middle| x \text{ is a real number}\right\}$ is a subgroup of the group $N_{2 \times 2} = \left\{\begin{pmatrix} a & b \\ c & d \end{pmatrix} \middle| ad - bc \neq 0\right\}$ under the operation of multiplication.

b. Other than the subgroup $\left\{\begin{pmatrix} 1 & 0 \\ 0 & 1 \end{pmatrix}\right\}$, are there any finite subgroups of $N_{2 \times 2}$? Verify your answer.

7. Let H_1 and H_2 be subgroups of a group G. Prove that $H_1 \cap H_2$ is also a subgroup.

8. Let G be a group and let $\mathcal{H}$ be a collection of any number of subgroups of G. Prove that the intersection of all the subgroups in $\mathcal{H}$ is a subgroup of G.

9. Is the union of two subgroups a subgroup?

10. Let G be any group and define

$$C = \{x \in G \mid x \circ g = g \circ x \text{ for all } g \text{ in } G\}.$$

Prove that C is a subgroup of G. C is called the **center** of G.

CYCLIC SUBGROUPS

We turn now to an important and useful type of subgroup—the cyclic subgroup.

DEFINITION

Let G be a group and let g be any element of G. We define

$\langle g \rangle = \{g^n \mid n \text{ is an integer}\}.$

That is, $\langle g \rangle = \{\ldots, g^{-2}, g^{-1}, g^0 = e, g^1, g^2, \ldots\}$, the set of all "powers" of g. We want to prove some theorems about $\langle g \rangle$, which turns out to be one of a very important class of subgroups.

THEOREM 5.4

If G is a group and $g \in G$, then $\langle g \rangle$ is a subgroup of G; that is, $\langle g \rangle < G$.
  Proof: We will use Theorem 5.2 to prove $\langle g \rangle < G$. First,

$\langle g \rangle \neq \emptyset$ since $g \in \langle g \rangle$.

Let x and y be in $\langle g \rangle$. Then $x = g^r$ and $y = g^s$ for some integers r and s. By the laws of exponents, $x \circ y^{-1} = g^r \circ (g^s)^{-1} = g^r \circ g^{-s} = g^{r-s}$. But $g^{r-s} \in \langle g \rangle$ since g^{r-s} is a power of g. Since x and y were any elements of H, $\langle g \rangle$ is a subgroup of G. ∎

With each element g of G, we have associated a subgroup $\langle g \rangle$. That is, for a different element $g_1 \in G$, we get a possibly different subgroup $\langle g_1 \rangle = \{g_1{}^n | n \in Z\}$.

DEFINITION

We call $\langle g \rangle$ the **cyclic subgroup** generated by g.

EXAMPLE 5.8

Consider the group of integers Z under addition and the element $2 \in Z$. Since the operation is addition, the second "power" of 2 is $2 + 2 = 4$; the nth power is

$$\underbrace{2 + 2 + \cdots + 2}_{n} = 2n.$$

Therefore

$$\langle 2 \rangle = \{2n | n \in Z\}$$

and we recognize $\langle 2 \rangle = E = $ the subgroup of all even integers.

EXAMPLE 5.9

In the group D_4 (Example 3.4) let us try to find $\langle r_1 \rangle$ the cyclic subgroup generated by "rotation through 90°."

$r_1{}^2 = r_1 \circ r_1 = r_2$

$r_1{}^3 = r_1{}^2 \circ r_1 = r_2 \circ r_1 = r_3$

$r_1{}^4 = r_1{}^3 \circ r_1 = r_3 \circ r_1 = I$

$r_1{}^5 = r_1{}^4 \circ r_1 = I \circ r_1 = r_1$

and so on.

Therefore $\langle r_1 \rangle = \{I, r_1, r_2, r_3\}$. Note that the pattern for positive powers of r_1 has been established above. Referring to the operation table for D_4, we see that $r_1{}^{-1} = r_3 \in \langle r_1 \rangle$, $(r_1{}^2)^{-1} = r_2{}^{-1} = r_2 \in \langle r_1 \rangle$, and so on; the set $\{I, r_1, r_2, r_3\}$ includes negative powers of r_1 as well as positive powers.

Computing $\langle r_2 \rangle$ and $\langle r_3 \rangle$ in a similar way, we see that $\langle r_2 \rangle = \{I, r_2\}$, while $\langle r_3 \rangle = \langle r_1 \rangle = \{I, r_1, r_2, r_3\}$. (Why?)

EXERCISES

1. a. Let Z be the additive group of integers. What is $\langle 3 \rangle$? $\langle 1 \rangle$? $\langle -1 \rangle$? Recall that the operation is addition.

   b. Find an explicit representation for $\langle 2 \rangle \cap \langle 3 \rangle$. (*Note:* $\langle 2 \rangle \cap \langle 3 \rangle$ is a subgroup by Exercise 7 of the last exercise set.)

c. Can you generalize the result in part b to find an explicit representation of $\langle n \rangle \cap \langle m \rangle$ for any positive integers n and m?
2. In the group S_3 (Example 3.5) find $\langle f_0 \rangle, \langle f_1 \rangle$, and $\langle f_3 \rangle$.
3. For $G = \{e, a, b, c\}$, the Klein four-group, find $\langle e \rangle, \langle a \rangle, \langle b \rangle$, and $\langle c \rangle$.
4. Consider the cyclic group G_6 (Example 3.3). Find $\langle g \rangle$ for each $g \in G_6$.
5. In the group $(M_{2 \times 2}, +)$ what is $\left\langle \begin{pmatrix} -1 & 0 \\ 0 & -1 \end{pmatrix} \right\rangle$?
6. In the group $(N_{2 \times 2}, \cdot)$, what is $\left\langle \begin{pmatrix} -1 & 0 \\ 0 & -1 \end{pmatrix} \right\rangle$?
*7. Let G be a group, $g \in G$. Let $\mathscr{H}$ be the collection of all those subgroups of G which contain g. Prove that $\langle g \rangle$ = the intersection of all the subgroups in $\mathscr{H}$. (Do the cases for finite and infinite G's separately.)

We have previously started with a group, taken an element in it, and generated a cyclic subgroup. There is a cyclic subgroup associated with every element of a group. We now try a slightly different approach—looking at the internal structure of the original group.

EXAMPLE 5.10
In Exercise 4 of the last set, you were asked to find all the cyclic subgroups of the group G_6. If the problem was done correctly, you found that

$$\langle 1 \rangle = \langle 5 \rangle = G_6$$

or that G_6 is a cyclic subgroup of itself. In view of this fact we make the following definition.

DEFINITION
If G is a group and if there is an element $g \in G$ such that $\langle g \rangle = G$, then we call G a **cyclic group generated by g.**
g is called a **generator** of G.

It is easy to see that the cyclic groups introduced in Example 3.3 are, in fact, cyclic by this definition. They are each generated by the element 1. The Klein four-group (Example 3.2), is not cyclic as the results of Exercise 3 show.

EXERCISES
1. Look at Examples 1.1–1.7 and 3.1–3.5 (Chapters 1 and 3) and decide which groups are cyclic groups. Find their generators.
2. If G is a cyclic group, prove that G is abelian. That is, prove that $x \circ y = y \circ x$ for every x and y in G.

DIVISION ALGORITHM

Before we investigate the structure of cyclic groups further, we will need an additional result, the division algorithm. The basic concept of the division algorithm is that for two positive integers s and t, if s is divided by t, there is a quotient q and a nonnegative remainder r which is less than t. This is the same concept that is taught to children when they first learn to divide. We are formalizing it and proving it for later use.

The proof of the division algorithm is done by mathematical induction. This is the third example of an inductive proof in this book. (Theorems 4.5 and 4.6 were also proved by induction.) Recall that to prove a statement is true for all positive integers, we first prove it is true for the integer 1 and then assume it is true for the integer k to prove that it is true for $k + 1$. Watch for the pattern.

THEOREM 5.5: DIVISION ALGORITHM

Let s and t be positive integers. Then there exists unique integers q and r with

$$s = qt + r, \quad \text{where } 0 \leq r < t.$$

Proof: We will do this proof by induction on s.

Let $s = 1$: If $t = 1$ also, then $1 = 1 \cdot (1) + 0$, so $q = 1$ and $r = 0$. If $t > 1$, then $1 = 0 \cdot t + 1$, so $q = 0$, $r = 1$. In either case, $0 \leq r < t$. Therefore the result is true if $s = 1$ for any value of t.

Let $s = k$: Choose any positive value of t and assume that there exist unique integers q and r with $k = qt + r$ where $0 \leq r < t$.

For $s = k + 1$, and the same arbitrary value of t, we want to find integers q_1 and r_1 with the desired property

$$k + 1 = qt + r + 1.$$

Note that since $r < t$, either $r + 1 < t$ or $r + 1 = t$.

Case 1: If $(r + 1) < t$, then the values $q_1 = q$ and $r_1 = r + 1$ give us the desired result

$$k + 1 = qt + r + 1 = q_1 t + r_1.$$

Case 2: If $r + 1 = t$, then

$$k + 1 = qt + t$$
$$= (q + 1)t + 0$$

and the values $q_1 = q + 1$ and $r = 0$ give us the result.

In either case, we have found integers q_1 and r_1 with $k + 1 = q_1 t + r_1$, where $0 \leq r_1 < t$. Since the theorem is true for $s = 1$, and given that it holds for $s = k$ implies that it is true for $s = k + 1$, for any positive integers s and t we can find integers q and r such that

$$s = qt + r, \quad \text{where } 0 \leq r < t.$$

We still have to prove that for fixed values of s and t, the integers q and r are unique.

Assume that $s = qt + r = q_1 t + r_1$, where $0 \leq r \leq r_1 < t$. Collecting terms we get $(q - q_1)t = r_1 - r$. But t divides $(q - q_1)t$, so it must also divide $r_1 - r$. But since both r and r_1 are between 0 and t, their difference is surely less than t. Therefore, for t to divide $r_1 - r$, we must have $r_1 - r = 0$, or $r_1 = r$. Finally $s = qt + r = q_1 t + r$ implies $qt = q_1 t$ or $q = q_1$.

For any pair of positive integers s and t, we have proved both existence and uniqueness of the desired integers q and r. ∎

We will be using the inductive method of proof many times in this book. It is an important mathematical tool and should be mastered as early as possible. A rereading of the proofs of Theorems 4.5 and 5.5 should help in this mastery.

EXAMPLES 5.11
For $s = 13, t = 5$,

$$13 = 2 \cdot 5 + 3 \quad (q = 2, r = 3).$$

For $s = 2, t = 7$,

$$2 = 0 \cdot 7 + 2 \quad (q = 0, r = 2).$$

For $s = 40, t = 5$,

$$40 = 8 \cdot 5 + 0 \quad (q = 8, r = 0).$$

For $s = 384, t = 125$,

$$384 = 3(125) + 9 \quad (q = 3, r = 9).$$

COROLLARY

Let s and t be positive integers. Then there exist integers q and r such that

$$-s = qt + r, \quad \text{where } 0 \leq r < t.$$

Note: This corollary extends the result of the theorem to all integers s, whether positive or negative. (Why does it hold for $s = 0$?) The extension for all values of t except zero is done in Exercise 2.

Proof: By Theorem 5.5, there exists a pair q_1 and r_1 such that $s = q_1 t + r_1$, where $0 \leq r_1 < t$. Multiplying by -1 gives $-s = -(q_1 t + r_1) = (-q_1)t - r_1$.

Case 1: If $r_1 = 0$, let $q = -q_1$ and $r = 0$. Then $-s = qt + 0$ is the desired result.

Case 2: If $r_1 \neq 0$, write the expression for $-s$ as

$$-s = (-q_1 t) - t + t - r_1 \quad \text{(adding } 0 = -t + t\text{)}$$
$$= (-q_1 t) - t + (t - r_1)$$
$$= (-q_1 - 1)t + (t - r_1).$$

Let $q = -q_1 - 1$ and let $r = t - r_1$. Now $-s = qt + r$, and since $0 < r_1 < t$, we have $0 < r = (t - r_1) < t$. The pair of integers, q and r, works. ■

EXAMPLE 5.12
For $s = 7$ and $t = 9$,

$$7 = 0 \cdot 9 + 7, \qquad q_1 = 0 \text{ and } r_1 = 7$$

so that $q = -q_1 - 1 = 0 - 1 = -1$ and $r = t - r_1 = 9 - 7 = 2$. We see that

$$-7 = (-1)(9) + 2.$$

Notation: If, for integers s and t, $t \neq 0$, we have the relation $s = qt + 0$, we say that t divides s and denote this situation as $t \mid s$.

EXERCISES

1. Show that Theorem 5.5 holds for $s = 0$ and any positive integer t.
2. Prove Theorem 5.5 for any integers s and t except $t = 0$. You may assume the cases that were proved in the text. Why do we exclude $t = 0$?
3. Find the integers q and r for the following pairs s and t, so $s = qt + r$, $0 \leq r < t$.
    a. $s = 3, t = 2$
    b. $s = 2, t = 3$
    c. $s = 34, t = 17$
    d. $s = -89, t = 23$
    e. $s = 0, t = 7$
    f. $s = 3, t = -11$
    g. $s = -4, t = -3$
4. Prove that for any integer n, exactly one of the integers $n, n - 1, n - 2, n - 3$ is divisible by 4.

We make our first use of the division algorithm to prove the following theorem.

THEOREM 5.6
Every subgroup of a cyclic group is cyclic.

Proof: Let G be a cyclic group, $G = \langle g \rangle$, and let H be a subgroup of G.
Case 1: $H = \{e\}$, then $H = \langle e \rangle$ which is cyclic.
Case 2: If $H \neq \{e\}$, then every element of H is of the form g^n and there are both positive and negative powers of g in H. Let k be the smallest

positive integer so that g^k is in H. We want to show that $H = \langle g^k \rangle$. Since H is closed under the operation, all powers of g^k are in H or $\langle g^k \rangle \subseteq H$. We need to show that $H \subseteq \langle g^k \rangle$.

Choose an element $g^m \in H$. We want to show that $m = k \cdot q$ for some integer q. By the division algorithm, there exist integers q and r with $m = qk + r, 0 \leq r < k$, and therefore

$$g^m = g^{qk+r} = g^{qk} \circ g^r.$$

We can write $g^r = g^m \circ (g^k)^{-q}$, the product of two elements in H, and therefore $g^r \in H$. Since k is the smallest positive integer with $g^k \in H$ and $0 \leq r < k$, then r must be 0. We have proved that $m = kq$, and $H \subseteq \langle g^k \rangle$. $H \subseteq \langle g^k \rangle$ and $\langle g^k \rangle \subseteq H$ implies $H = \langle g^k \rangle$ or H is cyclic. ∎

DEFINITION

If G is a group and $g \in G$, then the smallest positive integer n such that $g^n = e$ is called the **order** of g. If no such n exists, then g is said to be of **infinite order**.

In Chapter 2 we defined the order of a group G, $|G|$, as the number of elements in G. We will investigate the relationship between these two seemingly different definitions by looking at a cyclic group $\langle g \rangle$.

THEOREM 5.7

If G is a group and $g \in G$ and if n is the smallest positive power of g such that $g^n = e$ (the order of g is finite), then $\langle g \rangle = \{e, g, g^2, \ldots, g^{n-1}\}$, where this list contains distinct elements.

Proof: There are two parts to this proof. First, we have to prove that *any* power of g can be written as an element of the set $\{e, g, g^2, \ldots, g^{n-1}\}$. We also have to prove that these n elements are all distinct.

To prove the first part, let s be any integer, positive or negative. Then by the division algorithm, we can find integers q and r such that $s = qn + r$, and $0 \leq r < n$. Thus

$$\begin{aligned} g^s = g^{qn+r} &= g^{qn} \circ g^r \\ &= (g^n)^q \circ g^r \\ &= e^q \circ g^r \quad \text{(since } g^n = e\text{)} \\ &= e \circ g^r = g^r. \end{aligned}$$

Since $0 \leq r < n$, g^r is an element of $\{e = g^0, g^1, g^2, \ldots, g^{n-1}\}$ as we claimed. We have done the first part of the proof.

For the second, we will do a proof by contradiction. Assume that two of the elements of $\{e, g, g^2, \ldots, g^{n-1}\}$ are equal, say $g^k = g^l$, where $0 \leq k < l \leq n-1$. Then

$$g^{l-k} = g^l \circ g^{-k}$$
$$= g^l \circ (g^k)^{-1}$$
$$= g^l \circ (g^l)^{-1} \quad \text{(since } g^k = g^l\text{)}$$
$$= e.$$

However, since $0 < l - k < n$, this contradicts our assumption that n is the *least* positive integer such that $g^n = e$. Therefore $g^{l-k} \neq e$ and $g^l \neq g^k$. Thus, as claimed, we have the result that the subgroup generated by g, $\langle g \rangle = \{e, g, g^2, \ldots, g^{n-1}\}$ where all the elements are distinct. ∎

COROLLARY

If g has order n, then $\langle g \rangle$ has order n.

Proof: By Theorem 5.7, $\langle g \rangle = \{e, g, g^2, \ldots, g^{n-1}\}$ which has n elements. ∎

EXAMPLE 5.13

In D_4 look at the element r_1, rotation through 90°. As we saw in Example 5.9, $r_1^4 = I$, the identity, and 4 is the smallest positive integer for which this happens. The order of r_1 is 4.

The cyclic subgroup generated by r_1 is $\{I, r_1, r_1^2, r_1^3\} = \{I, r_1, r_2, r_3\}$, which is exactly what the theorem guarantees.

We still need to see what happens if the order of g is infinite.

THEOREM 5.8

If G is a group and $g \in G$ is a group element with infinite order, then if n and m are integers, $n \neq m$, $g^n \neq g^m$.

Proof: We note that, by assumption, $g^k \neq e$ for any positive integer k, and proceed to prove the theorem by contradiction.

Assume there exist distinct integers n and m such that $g^n = g^m$. Then

$$g^{n-m} = g^n \circ g^{-m}$$
$$= g^n \circ (g^m)^{-1}$$
$$= g^n \circ (g^n)^{-1} \quad \text{(since } g^n = g^m\text{)}$$
$$= e.$$

Also, $g^{m-n} = e$, by a similar argument. But either $n - m > 0$ or $m - n > 0$, which contradicts our assumption that g has infinite order. Therefore there are no integers n and m for which $g^n = g^m$. ∎

Subgroups and cyclic subgroups   51

COROLLARY

If g is a group element with infinite order, then the order of $\langle g \rangle$ is infinite.

Proof: $\langle g \rangle = \{\ldots, g^{-n}, \ldots, g^{-2}, g^{-1}, e, g^1, g^2, \ldots, g^n, \ldots\}$ and by Theorem 5.8, all of the elements are distinct; therefore $\langle g \rangle$ has infinite order. ∎

EXAMPLE 5.14

In the group $N_{2 \times 2} = \left\{ \begin{pmatrix} a & b \\ c & d \end{pmatrix} \middle| ad - bc \neq 0 \right\}$ under multiplication, consider the element $g = \begin{pmatrix} 1 & 1 \\ 0 & 1 \end{pmatrix}$ and look at $\langle g \rangle$.

$$g^1 = \begin{pmatrix} 1 & 1 \\ 0 & 1 \end{pmatrix}$$

$$g^2 = \begin{pmatrix} 1 & 1 \\ 0 & 1 \end{pmatrix}\begin{pmatrix} 1 & 1 \\ 0 & 1 \end{pmatrix} = \begin{pmatrix} 1 & 2 \\ 0 & 1 \end{pmatrix}$$

⋮

$$g^n = \begin{pmatrix} 1 & 1 \\ 0 & 1 \end{pmatrix}^n = \begin{pmatrix} 1 & n \\ 0 & 1 \end{pmatrix} \quad \text{(Can you prove this?)}$$

⋮

Also,

$$g^{-1} = (g^1)^{-1} = \begin{pmatrix} 1 & -1 \\ 0 & 1 \end{pmatrix}$$

⋮

$$g^{-n} = (g^n)^{-1} = \begin{pmatrix} 1 & -n \\ 0 & 1 \end{pmatrix}.$$

⋮

Therefore the order of g is infinite and

$$\langle g \rangle = \left\{ \begin{pmatrix} 1 & n \\ 0 & 1 \end{pmatrix} \middle| n \text{ is an integer} \right\}$$

which has an infinite number of elements and has, therefore, infinite order.

We repeat and emphasize some of the more important work we have just done. If $(G, \circ)$ is a group, then the **order of G** is the number of elements in G, written $|G|$. If $g \in G$, then the **order of g** is the least positive integer n such that $g^n = e$. If no such integer exists, we say that g has infinite order. In either case, we have seen that *the order of the element g is always equal to the order of $\langle g \rangle$, the cyclic subgroup generated by g.*

EXERCISES

1. Find the orders of each of the elements of the Klein four-group.
2. a. What is the order of $\begin{pmatrix} -1 & 0 \\ 0 & -1 \end{pmatrix}$ in the group $(M_{2\times 2}, +)$?

   b. What is the order of $\begin{pmatrix} -1 & 0 \\ 0 & -1 \end{pmatrix}$ in the group $(N_{2\times 2}, \cdot)$?
3. Find the order of every element of the cyclic groups G_5, G_6, and G_{12} and find all generators of these groups. Can you generalize these results to determine the generators of G_n?
4. Show that the order of g^{-1} is the same as the order of g for any element g of a group G.
5. Let g be an element of order mn, where m and n are integers. Show that the order of g^m is n.
6. Let G be a group with no proper subgroups (G and $\{e\}$ are the only subgroups).

    a. Show G is cyclic. (*Hint:* If $g \in G$, what is $\langle g \rangle$?)

    b. Show $|G|$ is prime. (Use Exercise 5.)

Equivalence relations, cosets, normal subgroups, and Lagrange's theorem

This chapter is ultimately devoted to proving a very beautiful and important result—Lagrange's Theorem. However, in developing the machinery needed to prove this theorem, we investigate three other useful concepts.

EQUIVALENCE RELATIONS

This concept applies to other sets as well as to groups. Although we will first be using it to prove facts about groups, we will develop equivalence relations in the most general sense so that the concept can be applied in other situations as well.

DEFINITION

If A and B are sets, then $A \times B = \{(a, b) | a \in A \text{ and } b \in B\}$ is called the **cartesian product** of A and B.

EXAMPLES 6.1

1. If $A = \{a, b, c\}$ and $B = \{1, 2, 3, 4\}$, then

$A \times B = \{(a, 1)(a, 2)(a, 3)(a, 4)(b, 1)(b, 2)(b, 3)(b, 4)(c, 1)(c, 2)(c, 3)(c, 4)\}$

while

$B \times A = \{(1, a)(1, b)(1, c)(2, a)(2, b)(2, c)(3, a)(3, b)(3, c)(4, a)(4, b)(4, c)\}.$

We see that $A \times B \neq B \times A$.

2. If $Z = \{n | n \text{ is an integer}\}$, we form the cartesian product $Z \times Z = \{(n, m) | n \text{ and } m \text{ are integers}\}$. The cartesian product $Z \times Z$ has a graphical representation as the lattice points in the plane.

3. If $(G_1, \circ)$ and $(G_2, *)$ are groups, then $G_1 \times G_2 = \{(g_1, g_2) | g_1 \in G_1$ and $g_2 \in G_2\}$. We define an operation • on $G_1 \times G_2$ by

$$(g_1, g_2) \bullet (h_1, h_2) = (g_1 \circ h_1, g_2 * h_2).$$

In Exercise 1 you are asked to show that $(G_1 \times G_2, \bullet)$ is a group. $(G_1 \times G_2, \bullet)$ is called the **external direct product** of G_1 and G_2.

DEFINITION
Any subset $r \subset A \times A$ is called a **relation** on a set A. If $(a, b) \in r$, then we write $a \, r \, b$, read a **is related to** b.

If we give the relation (subset of $A \times A$) a different name, such as $\sim$, then $(a, b) \in \sim$ would be written as $a \sim b$. Basically, a relation is a subset of a cartesian product and is therefore a set of ordered pairs. However the notation $a \, r \, b$ is used for $(a, b) \in r$ for convenience. Example 6.2 will show that it is even more familiar in some relations.

EXAMPLES 6.2
1. If $A = \{a, b, c\}$, $A \times A = \{(a, a)(a, b)(a, c)(b, a)(b, b)(b, c)(c, a)(c, b)(c, c)\}$. Any subset of $A \times A$ is a relation on A. For example, $r = \{(a, b)(a, c)(b, c)\}$ is a relation with $a \, r \, b$, $a \, r \, c$, and $b \, r \, c$.
2. In $\mathbb{R}$, the set of real numbers, we define a relation $<$ on $\mathbb{R}$ by

$$< \; = \{(x, y) | x \text{ is less than } y\}.$$

If the pair (x, y) is in $<$, we write $x < y$. For example, $3 < 5$, $-2 < 4$, $0 < 3$, and so on. The relation $<$ has a graphical representation on the plane.

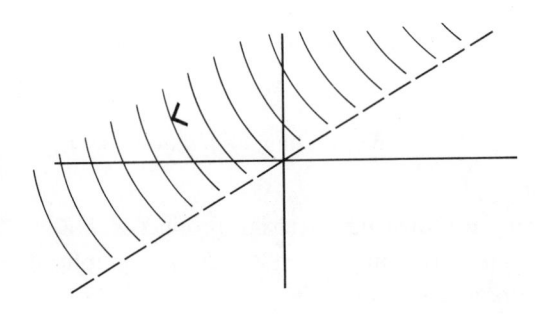

Cosets, normal subgroups, and Lagrange's theorem

3. For any set A, the set $A \times A$ is itself a relation on A in which every element of A is related to every other element. Also, $\emptyset$, the empty set, is a relation on A in which no element of A is related to any other element.

DEFINITION

An **equivalence relation** r on a set A is a relation on A which satisfies all of the following properties:
1. a r a for every $a \in A$ (reflexive).
2. If a r b, then b r a (symmetric).
3. If a r b and b r c, then a r c (transitive).

EXAMPLES 6.3

1. Equality is the most famous and widely used equivalence relation. We talk of equality of integers, of real numbers, of matrices, of sets, and so on. On the set $B = \{1, 2, 3\}$, the relation of equality is

$$= \; = \{(1, 1)(2, 2)(3, 3)\}.$$

We can see immediately that $=$ is an equivalence relation on any set since
a. It is reflexive: $a = a$ for every a in the set.
b. It is symmetric: $a = b \Rightarrow b = a$.
c. It is transitive: $a = b$ and $b = c \Rightarrow a = c$.
In fact, these are properties of equality that we use in almost any algebraic problem.

2. We define a relation $\equiv_5$ on the integers:

$$\equiv_5 \; = \{(a, b) | a - b \text{ is divisible by } 5\}.$$

$a \equiv_5 b$ or $a \equiv b \pmod{5}$ is read a is congruent to b mod 5 and means $5|(a - b)$. We will prove that $\equiv_5$ is an equivalence relation on the set of integers Z.

a. To show that $\equiv_5$ is reflexive, choose any integer $a \in Z$. $a - a = 0$ and $5|0$ since $0 = 5 \cdot 0$. Therefore, for any $a \in Z$, $a \equiv_5 a$ and $\equiv_5$ is reflexive.

b. If $a, b \in Z$ and $a \equiv_5 b$, then $5|(a - b)$ or $a - b = 5k$ for some k in Z. We can write

$$b - a = -(a - b)$$
$$= -(5 \cdot k)$$
$$= 5(-k)$$

so that $5|(b - a)$ or $b \equiv_5 a$. Since $a \equiv_5 b$ implies $b \equiv_5 a$, $\equiv_5$ is symmetric.

c. If $a, b, c \in Z$ with $a \equiv_5 b$ and $b \equiv_5 c$, then there are integers k and l so that $a - b = 5k$ and $b - c = 5l$. We want to show that $a - c$ is divisible by 5.

$$\begin{aligned} a - c &= (a - b) + (b - c) \\ &= 5k + 5l \\ &= 5(k + l). \end{aligned}$$

Since $k + l$ is an integer, $a - c$ is divisible by 5. Therefore $a \equiv_5 b$ and $b \equiv_5 c$ implies $a \equiv_5 c$ or $\equiv_5$ is transitive. $\equiv_5$ is an equivalence relation since it satisfies all three of the properties.

3. If G is a group and $H < G$, consider the relation r defined on G by

$$r = \{(x, y) | x \circ y^{-1} \in H\}.$$

We will show that r is an equivalence relation.

a. Reflexive: For every $x \in G$, $x \circ x^{-1} = e \in H$, since H is a subgroup. Therefore x r x for every $x \in G$.

b. Symmetric: If $x \circ y^{-1} \in H$, (x r y), then, since H is a subgroup, $(x \circ y^{-1})^{-1} \in H$. By Theorem 4.6 $(x \circ y^{-1})^{-1} = (y^{-1})^{-1} \circ x^{-1} = y \circ x^{-1}$. Therefore $y \circ x^{-1} \in H$ or y r x.

c. Transitive: If $x \circ y^{-1} \in H$ and $y \circ z^{-1} \in H$, (x r y and y r z), then, since H is closed under $\circ$,

$$(x \circ y^{-1}) \circ (y \circ z^{-1}) = x \circ z^{-1} \in H$$

or x r z.

DEFINITION

If A is a set and r is an equivalence relation on A, then for each element $a \in A$, we define the **equivalence class** of the element a as the set $[a] = \{x | x \text{ r } a\}$.

The equivalence class of an element $a \in A$ is the set of all elements that are related to a under r.

EXAMPLES 6.4

We will find the equivalence classes for each of the equivalence relations in Examples 6.3.

1. For equality on the set $B = \{1, 2, 3\}$.

The equivalence class of $1 = [1] = \{x \in B | x = 1\} = \{1\}$.
The equivalence class of $2 = [2] = \{x | x = 2\} = \{2\}$.
The equivalence class of $3 = [3] = \{x | x = 3\} = \{3\}$.

Cosets, normal subgroups, and Lagrange's theorem 57

2. For the relation of congruence modulo 5, we need to find an equivalence class for each integer.

$[0] = \{x | x \equiv_5 0\} = \{x | 5 \text{ divides } (x - 0)\} = \{5n | n \in Z\}$.
$[0] = \{\ldots, -10, -5, 0, 5, 10, 15, \ldots\}$.

Note that
$[0] = [5] = [10] = [15] = \cdots$
$\quad = [-5] = [-10] = \cdots$.

Similarly

$[1] = \{x | x \equiv_5 1\} = \{x | 5 \text{ divides } (x - 1)\} = \{5n + 1 | n \in Z\}$.
(If $x \in [1]$, then $x - 1 = 5n$ for some $n \in Z$ or $x = 5n + 1$.)
$[1] = \{\ldots, -9, -4, 1, 6, 11, 16, \ldots\}$.

Notice that
$[1] = [6] = [11] = \cdots$
$\quad = [-4] = [-9] = \cdots$.

For example,
$[6] = \{x | x \equiv_5 6\}$
$\quad = \{x | x - 6 = 5l, l \in Z\}$
$\quad = \{x | x = 5l + 6 = 5(l + 1) + 1 = 5n + 1, n \in Z\}$.

In a similar fashion, we see that

$[2] = \{x | x = 5k + 2, k \in Z\} = \{\ldots, -8, -3, 2, 7, 12, \ldots\}$
$\cdots = [-8] = [-3] = [2] = [7] = [12] = \cdots$.
$[3] = \{x | x = 5k + 3, k \in Z\} = \{\ldots, -7, -2, 3, 8, 13, \ldots\}$
$\cdots = [-7] = [-2] = [3] = [8] = [13] = \cdots$.
$[4] = \{x | x = 5k + 4, k \in Z\} = \{\ldots, -6, -1, 4, 9, 14, \ldots\}$
$\cdots = [-6] = [-1] = [4] = [9] = [14] = \cdots$.

There are only 5 distinct equivalence classes for all the integers under the relation $\equiv_5$. (Why?)

3. In Example 6.3.3, G is a group, H is a subgroup of G, and $x \, r \, y$ means $x \circ y^{-1} \in H$. We need to find an equivalence class for every element $g \in G$.

Case 1: For $h \in H$, we will find $[h]$. Consider an element $x \in G$ such that $x \, r \, h$; this implies that $x \circ h^{-1} \in H$ or, for some $h_1 \in H$, $x \circ h^{-1} = h_1$ or $x = h_1 \circ h$, a product of two elements in H. Therefore $x \in [h] \Rightarrow x \in H$

and $[h] \subseteq H$. Also, if $x \in H$, $x \circ h^{-1} \in H$ since H is closed, and $x \in [h]$ or $H \subseteq [h]$. Therefore $[h] = H$.

Case 2: For $g \in G$, $g \notin H$, we will find $[g]$. If $x \in [g]$, $x \text{ r } g$ and then $x \circ g^{-1} \in H$. For some h_1 in H, $x \circ g^{-1} = h_1$ or $x = h_1 \circ g$. We see that $x \in [g] \Rightarrow x \in \{h \circ g | h \in H\}$ or $[g] \subseteq \{h \circ g | h \in H\}$. In addition, if $x \in \{h \circ g | h \in H\}$, then for some $h_2 \in H$, $x = h_2 \circ g$. Therefore $x \circ g^{-1} = h_2 \in H$ or $x \text{ r } g$. We now have $[g] \subseteq \{h \circ g | h \in H\}$ and $\{h \circ g | h \in H\} \subseteq [g]$ so that $[g] = \{h \circ g | h \in H\}$.

In Case 1, $[h] = H$ for $h \in H$. However, notice that the set $H \circ h = \{h_1 \circ h | h_1 \in H\} = H$. (Why?) Therefore, for any $g \in G$ (whether g is in H or not), $[g] = \{h \circ g | h \in H\}$.

In Examples 6.4.1 and 6.4.2 we can see that each element of the underlying set is in an equivalence class and that distinct equivalence classes are disjoint. We will now prove that this is a general result.

THEOREM 6.1

If A is a set and r is an equivalence relation on A, then
1. every element of A belongs to at least one equivalence class of r and
2. if $a, b \in A$ and $[a] \cap [b] \neq \emptyset$, then $[a] = [b]$.

Proof of 1: Since r is an equivalence relation, r is reflexive; that is, for every $a \in A$, $a \text{ r } a$ and $a \in [a]$ or each element belongs to its own equivalence class.

Proof of 2: Part 2 states that equivalence classes are either disjoint or identical. Assume that a and b are elements of A with $[a] \cap [b] \neq \emptyset$. We will show that $[a] = [b]$. $[a] \cap [b] \neq \emptyset$ implies that there is an x in $[a] \cap [b]$. Therefore $x \in [a]$ and $x \in [b]$ or $x \text{ r } a$ and $x \text{ r } b$. Since r is an equivalence relation and therefore symmetric, we have $a \text{ r } x$ and $x \text{ r } b$. By the transitive property of r we have

$a \text{ r } b$.

Now, for any $y \in [a]$, $y \text{ r } a$ and $a \text{ r } b$ implies $y \text{ r } b$ or $y \in [b]$. Therefore $[a] \subseteq [b]$.

To prove that $[b] \subseteq [a]$, we note that $a \text{ r } b$ implies $b \text{ r } a$. Now, for any $z \in [b]$, $z \text{ r } b$ and $b \text{ r } a$ implies $z \text{ r } a$ and $z \in [a]$.

Since $[a] \subseteq [b]$ and $[b] \subseteq [a]$, $[a] = [b]$. ∎

Since distinct equivalence classes are disjoint, any element of an equivalence class can be used as a representative of its class. In Example 6.4.2 the equivalence class $\{\ldots, -3, 2, 7, 12, \ldots\}$ could be designated by each of the elements $[-3]$, $[2]$, $[497]$, $[-3038]$ or any other member of the class.

Cosets, normal subgroups, and Lagrange's theorem

DEFINITION

If A is a set, then a class of subsets of A, $\{A_1, A_2, A_3, \ldots\}$ is called a **partition** of A if
1. $A = A_1 \cup A_2 \cup A_3 \cup \cdots$ and
2. $A_i \cap A_j = \emptyset$, if $i \neq j$.

(*Note:* It is conceivable that a partition contains more than a countable number of elements, but, for notational convenience, we will proceed as if every partition is countable. The results are the same.)

EXAMPLES 6.5

1. If $A = \{a, b, c, d\}$, the class $\{\{a, c\} \{b\} \{d\}\}$ is a partition of A. $\{\{a\} \{b\} \{c\} \{d\}\}$ is another partition.

2. The sets $E = \{x \mid x \text{ is an even integer}\}$ and $0 = \{x \mid x \text{ is an odd integer}\}$ form a partition, $\{E, 0\}$, of the set Z of integers.

3. Theorem 6.1 guarantees that the equivalence classes associated with an equivalence relation form a partition of the underlying set. Examples 6.4.1, 6.4.2, 6.4.3 give us additional examples of partitions of a set.

We have seen that any equivalence relation induces a partition on the underlying set. We now prove the converse.

THEOREM 6.2

If A is a set with a partition $\{A_1, A_2, A_3, \ldots\}$, then there exists an equivalence relation r on A so that the equivalence classes are the sets of the partition.

Partial Proof: We define r as follows: For elements a and b in A, a r b if and only if a and b belong to the same set of the partition. That is, a r b implies that there is a set $A_i \in \{A_1, A_2, A_3, \ldots\}$ with $a \in A_i$ and $b \in A_i$.

We still need to show that (1) r is an equivalence relation and (2) the equivalence classes of r are exactly the sets of the partition. This will be done in the exercises.

EXAMPLES 6.6

1. Let $A = \{a, b, c, d\}$.

 a. $\{\{a\} \{b\} \{c\} \{d\}\}$ is a partition; equality is the relation induced. That is, x and y belong to the same equivalence class (set of the partition) if and only if $x = y$.

 b. $\{\{a, b\} \{c\} \{d\}\}$ is also a partition. The equivalence relation r induced by this partition is

$\{(a, a)(b, b)(a, b)(b, a)(c, c)(d, d)\}$.

2. The sets $E = \{x \mid x \text{ is even}\}$ and $0 = \{x \mid x \text{ is odd}\}$ form a partition of Z. The induced equivalence relation could be defined in a variety of ways.

One definition is

x r y means $x + y$ is divisible by 2.

We will finish this subsection on equivalence relations by drawing an important parallel between one of our examples here and a group we have studied previously. Let us investigate Examples 6.3.2 and 6.4.2 more thoroughly.

In Example 6.3.2, we showed that $\equiv_5$, defined on the set of integers by $a \equiv_5 b$ means $5|(a - b)$ is an equivalence relation. In Example 6.4.2, we found the 5 disjoint equivalence classes. We now define an operation on the equivalence classes by

$[a] + [b] = [a + b]$.

For example, $[1] + [2] = [3]$ or

$\{\ldots, -4, 1, 6, 11, \ldots\} + \{\ldots, -3, 2, 7, \ldots\} = \{\ldots, -2, 3, 8, 13, \ldots\}$.

Also, $[4] + [3] = [7] = [2]$.

We need to check to see that we get the same result if different equivalence class representatives are chosen. For example, $[4] = [9]$ and $[3] = [-12]$, $[9] + [-12] = [-3]$ but $[9] + [-12] = [-3] = [2] = [4] + [3]$, so that the sum of these two classes does not appear to depend on the choice of representatives. We will try to prove that, in general, this operation is well defined.

For two integers a and b, we defined $[a] + [b] = [a + b]$. Let us choose other representatives of the classes $[a]$ and $[b]$. Suppose $c \in [a]$ and $d \in [b]$, $[c] + [d] = [c + d]$, but does $[c + d] = [a + b]$? If $c \in [a]$, then $c - a = 5k$ for some integer k, or $c = a + 5k$. Similarly $d = b + 5l$ for some $l \in Z$. Now $c + d = a + b + 5(k + l)$ or $(c + d) \in [a + b]$. Theorem 6.1 tells us that these classes are identical or $[a + b] = [c + d]$. Therefore the sum of two equivalence classes does not depend on the representatives chosen and the operation is well defined.

If, for each equivalence class, we choose the single member between 0 and 4, inclusive, as the representative of that class, we obtain the following table:

+	[0]	[1]	[2]	[3]	[4]
[0]	[0]	[1]	[2]	[3]	[4]
[1]	[1]	[2]	[3]	[4]	[0]
[2]	[2]	[3]	[4]	[0]	[1]
[3]	[3]	[4]	[0]	[1]	[2]
[4]	[4]	[0]	[1]	[2]	[3]

Cosets, normal subgroups, and Lagrange's theorem

Comparing this table to the one accompanying Example 3.3 we see a marked similarity. In fact, with the omission of the braces, [ ], they are identical. The equivalence relation $\equiv_5$ provides us with another approach to the cyclic group G_5.

We take this approach one step further and define a multiplication operation on the set of equivalence classes of $\equiv_5$ on Z.

$$[a] \cdot [b] = [ab].$$

For example, $[3] \cdot [2] = [6] = [1]$ and $[4] \cdot [1] = [4]$. The proof that this operation is well defined is left as an exercise.

Again using the equivalence class representative between 0 and 4 inclusive, we construct the following table:

·	[0]	[1]	[2]	[3]	[4]
[0]	[0]	[0]	[0]	[0]	[0]
[1]	[0]	[1]	[2]	[3]	[4]
[2]	[0]	[2]	[4]	[1]	[3]
[3]	[0]	[3]	[1]	[4]	[2]
[4]	[0]	[4]	[3]	[2]	[1]

Obviously this is *not* a group since [0] has no inverse, but if we delete the first row and column of the table, we obtain

·	[1]	[2]	[3]	[4]
[1]	[1]	[2]	[3]	[4]
[2]	[2]	[4]	[1]	[3]
[3]	[3]	[1]	[4]	[2]
[4]	[4]	[3]	[2]	[1]

After checking properties 1, 2, and 3 of the definition, we can see that $\{[1] [2] [3] [4]\}$ is a group under the operation.

EXERCISES

1. If $(G_1, \circ)$ and $(G_2, *)$ are groups, show that the external direct product $G_1 \times G_2$ is a group under the operation defined in the text.
2. Which of the following relations are equivalence relations on the set of real numbers? Prove your results.
   a. x r y if xy is rational.
   b. x r y if $x \leq y$.
   c. x r y if xy is real.
   d. x r y if x/y is real.
   e. x r y if $xy = 0$.
   f. x r y if $x - y$ is an integer.
   g. x r y if $x - y = 0$.

3. Find the equivalence classes for each equivalence relation in Exercise 2.
4. a. If G is a group and H is a subgroup of G, prove that the relation on G defined by "$x \, r \, y$ if and only if $x^{-1} \circ y \in H$" is an equivalence relation.
   b. Find the associated equivalence classes.
5. Prove that a relation r on a set A is an equivalence relation if it is reflexive and satisfies the property $a \, r \, b$ and $b \, r \, c$ implies $c \, r \, a$.
6. Which of the following are partitions on the set of integers? Prove your results.
   a. $\{P, N\}$ where $P = \{x \mid x \text{ is positive}\}$ and $N = \{x \mid x \text{ is negative}\}$.
   b. $\{\text{II, III}, N\}$ where $\text{II} = \{x \mid x \text{ is divisible by 2}\}$, $\text{III} = \{x \mid x \text{ is divisible by 3}\}$, $N = \{x \mid x \text{ is not divisible by either 2 or 3}\}$.
   c. $\{Q, C\}$ where $Q = \{x \mid x \text{ is prime}\}$ and $C = \{x \mid x \text{ is composite}\}$.
   d. $\{R_0, R_1, R_2\}$ where $R_0 = \{3n \mid n \in Z\}$, $R_1 = \{3n + 1 \mid n \in Z\}$, and $R_2 = \{3n + 2 \mid n \in Z\}$.
7. Complete the proof of Theorem 6.2.
8. Show that multiplication on the equivalence classes of $\equiv_5$ is well defined.
9. a. Let n be a positive integer. Define a relation $\equiv_n$ on the integers by $a \equiv_n b$ if and only if $n \mid (a - b)$. Prove that $\equiv_n$ is an equivalence relation.
   b. Find the equivalence classes of $\equiv_n$.
   c. Show that the operation $+$ on the equivalence classes for $\equiv_n$, defined by $[a] + [b] = [a + b]$, is well defined.
   d. Construct a table for the set of equivalence classes for $\equiv_n$ under the operation of addition as defined in part c and show that this system is a group.
   e. Show that the operation $\cdot$ on the equivalence classes of $\equiv_n$, defined by $[a] \cdot [b] = [ab]$ is well defined.
   f. Can you determine for what values of n the set of all equivalence classes of $\equiv_n$ except $[0]$ form a group under the operation in part e?

COSETS

Throughout this section, G will be a group and H will be a subgroup of G, $H < G$.

DEFINITION

For $g \in G$, the set $H \circ g = \{h \circ g \mid h \in H\}$ is called the **right coset of the subgroup H in G determined by g.**

Note that for every element g in G, we get a right coset $H \circ g$. For every $g \in G$, we can also define the set $g \circ H = \{g \circ h \mid h \in H\}$, which is called **the left coset of H in G determined by g.** $H \circ g$ need not equal $g \circ H$ if the group G is not abelian. The results for left cosets are essentially the same as those for right cosets. We will do the results for right cosets and leave those for left cosets as an exercise.

Cosets, normal subgroups, and Lagrange's theorem

EXAMPLES 6.7
1. As a specific example, if G is the Klein four-group $G = \{e, a, b, c\}$ and $H = \{e, b\}$, then the four right cosets of H in G are

$H \circ e = \{e \circ e, b \circ e\} = \{e, b\}$
$H \circ a = \{e \circ a, b \circ a\} = \{a, c\}$
$H \circ b = \{e \circ b, b \circ b\} = \{b, e\} = \{e, b\} = H \circ e$
$H \circ c = \{e \circ c, b \circ c\} = \{c, a\} = \{a, c\} = H \circ a.$

2. If we take $S_3 = \{f_0, f_1, f_2, f_3, f_4, f_5\}$ (Example 3.5) and the subgroup $H = \{f_0, f_3, f_4\}$, the right cosets are

$H \circ f_0 = \{f_0, f_3, f_4\} = H \circ f_3 = H \circ f_4$
$H \circ f_1 = \{f_1, f_2, f_5\} = H \circ f_2 = H \circ f_5.$

Note that the coset $H \circ g$ need not be a subgroup of G. Note also that in these examples, at least, the cosets of H in G form a partition of G and by Theorem 6.2 there is an associated equivalence relation. We will prove that this is a general result in Theorem 6.6.

THEOREM 6.3
The number of elements in the right coset $H \circ g = |H|$, the order of H.

Proof: For every $h \in H$, we get an element $h \circ g \in H \circ g$. Using the cancellation laws we can prove that all the elements of $H \circ g$ are distinct: If $h, h' \in H$, then $h \circ g = h' \circ g$ if and only if $h = h'$. Therefore the correspondence $h \to h \circ g$ between elements of H and elements of $H \circ g$ is a one-to-one correspondence and the sets H and $H \circ g$ have the same number of elements. ∎

Let us consider a right coset $H \circ g$ of a subgroup H in an arbitrary group G. This is just a set of products $\{h \circ g | h \in H\}$. If we choose a different element of G, say g_1, then g_1 also determines a right coset $\{h \circ g_1 | h \in H\}$. Can it happen that these two right cosets have any elements in common, that is, have a nonempty intersection? The examples we have done show that it is possible. Under what specific conditions can it happen?

THEOREM 6.4
If two right cosets $H \circ g$ and $H \circ g_1$ have at least one element in common, then $g_1 = h \circ g$ for some element $h \in H$.

Proof: If two right cosets $H \circ g$ and $H \circ g_1$ have at least one element in common, then $(H \circ g) \cap (H \circ g_1) \neq \emptyset$. That is, some element of the set

$H \circ g_1$ is in $H \circ g$. Therefore, for some elements h_1 and h_2 in H,

$$h_1 \circ g_1 = h_2 \circ g.$$

Rearranging, we get

$$g_1 = (h_1^{-1} \circ h) \circ g.$$

Since H is a subgroup $h_1^{-1} \in H$ and the product $h_1^{-1} \circ h_2$ of elements of H are again in H, we can write $h_1^{-1} \circ h_2 = h$ for some $h \in H$ and $g_1 = h \circ g$. ∎

THEOREM 6.5
If $H < G$, and if $H \circ g$ and $H \circ g_1$ have at least one element in common, then

$$H \circ g_1 = H \circ g.$$

Proof: Assume $(H \circ g) \cap (H \circ g_1) \neq \emptyset$. By Theorem 6.4 we can write $g_1 = h \circ g$ for some $h \in H$ or $g_1 \in H \circ g$. We will show that $H \circ g_1 \subseteq H \circ g$ and vice versa.

Now if $h_1 \circ g_1$ is any element of $H \circ g_1$, we can write $h_1 \circ g_1 = h_1 \circ (h \circ g)$ since $g_1 = h \circ g$ or

$$h_1 \circ g_1 = h_2 \circ g$$

for $h_2 = h_1 \circ h \in H$. This implies that $h_1 \circ g_1 \in H \circ g$ or $H \circ g_1 \subseteq H \circ g$.

To show set containment in the other direction, we first note that $g_1 = h \circ g$ implies that $g = h^{-1} \circ g_1$ or that $g \in H \circ g_1$.

Now choose any element $h_3 \circ g \in H \circ g$. We can write

$$h_3 \circ g = h_3 \circ (h^{-1} \circ g_1) \quad \text{(since } g = h^{-1} \circ g_1\text{)}$$
$$= h_4 \circ g_1 \quad \text{(since } h_3 \circ h^{-1} = h_4\text{)}$$

for some $h_4 \in H$. Therefore $h_3 \circ g \in H \circ g_1$ or $H \circ g \subseteq H \circ g_1$.

Since $H \circ g \subseteq H \circ g_1$ and $H \circ g_1 \subseteq H \circ g$, we have proved that $H \circ g_1 = H \circ g$. ∎

We repeat this result in order to emphasize it. What it says is that two right cosets $H \circ g$ and $H \circ g_1$ are either disjoint or equal. There can be no overlap between two right cosets of a subgroup in a group unless the cosets are the same. As an illustration, look at the cosets in the examples above. There you can see that whenever two cosets are not disjoint, they are the same. Theorem 6.5 says that this situation is true in general.

Cosets, normal subgroups, and Lagrange's theorem

THEOREM 6.6

The class of all right cosets of H in G forms a partition of G.

Proof: In Theorem 6.5 we have proved that distinct cosets are disjoint. (We proved that if they are not disjoint, then they are not distinct, which is the same thing logically.)

We need to prove that G equals the union of all the cosets or, equivalently, that every element of G belongs to one of the cosets. For $g \in G$, we know that $e \in H$, since H is a subgroup, so that $g = e \circ g \in H \circ g$. That is, for every $g \in G$, g belongs to its own coset $H \circ g$. ∎

By Theorem 6.2 we know that there is an equivalence relation on G that produces the cosets $H \circ g$ as the associated equivalence classes. We define $g_2 \text{ r } g_1$ if and only if $H \circ g_1 = H \circ g_2$. (The proof of Theorem 6.2 that you did as an exercise proves that this is an equivalence relation with the cosets as the equivalence classes.)

$H \circ g_1 = H \circ g_2$ implies that for every h_1 in H there exists an element h_2 in H so that $h_1 \circ g_1 = h_2 \circ g_2$ or, rearranging,

$h_2^{-1} \circ h_1 = g_2 \circ g_1^{-1}$.

Since $h_2^{-1} \circ h_1 \in H$, this implies that $g_2 \circ g_1^{-1} \in H$ also. That is, $g_2 \text{ r } g_1$ means that $g_2 \circ g_1^{-1} \in H$. This is the equivalence relation that we worked with in Examples 6.3.3 and 6.4.3.

In view of this fact, we suggest that there is a much easier approach to cosets. It does not give as much intuitive insight as the work we have done here, but it is faster and it does tie things together.

Alternate development of cosets

Consider a group G and a subgroup $H < G$. We define a relation on G by

$x \text{ r } y$ means $x \circ y^{-1} \in H$.

THEOREM 6.7

r is an equivalence relation on G.

Proof: See Example 6.3.3.

THEOREM 6.8

For $g \in G$, the equivalence class of g with respect to the relation r is the set $\{h \circ g \mid h \in H\}$.

Proof: See Example 6.4.3.

DEFINITION

The equivalence class of g with respect to the equivalence relation r is called a **right coset** and is denoted by the symbol $H \circ g$.

EXERCISES

1. Look at the group of integers $(Z, +)$ and let $(E, +)$ be the subgroup of even integers. Find all the distinct right and left cosets.
2. Let $G = \{e, a, b, c\}$ be the Klein four-group; let $G_1 = \{e, a\}$, a subgroup of G. What are the distinct right cosets of G_1 in G?
3. Let $G_6 = \{0, 1, 2, 3, 4, 5\}$ be the cyclic group of order 6 and let $H = \{0, 2, 4\}$. What are the right cosets determined by 2? 3? 5?
4. Consider the group S_3 defined in Chapter 3. Find a subgroup H of order 3 and a subgroup H' of order 2. For each $f_i \in S_3$, $i = 0, 1, \ldots, 5$, compute explicitly the cosets $H \circ f_i$, $f_i \circ H$, $H' \circ f_i$, $f_i \circ H'$. Is it necessary that $H' \circ f = f \circ H'$?
5. Find all right and left cosets of the subgroups $\{I, r_1, r_2, r_3\}$ and $\{I, v\}$ in the group D_4 (Example 3.4).
6. Prove that if H is a subgroup of G, then there are as many left cosets of H in G as there are right cosets of H in G.
7. Formulate and prove results for left cosets that are analogous to the ones for right cosets proved in the text.

NORMAL SUBGROUPS

In our investigations of cosets, we have run across the question of equality of the right and left cosets $H \circ g$ and $g \circ H$. Both cosets have the same number of elements (by Theorem 6.3 and its analog for left cosets) and g belongs to both $H \circ g$ and $g \circ H$ since, for $e \in H$, $g = e \circ g = g \circ e$. But there is no guarantee that the rest of the elements are equal. Exercise 4 should have illustrated that result. $H \circ g$ does not necessarily equal $g \circ H$, but it is possible.

For example, the subgroup $\{f_0, f_3, f_4\}$ of S_3 has the property that all right cosets are left cosets. Any two-element subgroup of S_3 does not have that property.

This concept is so important that we give it a special name since we will be using it often in the next sections.

DEFINITION

If G is a group and N is a subgroup of G such that for every $g \in G$, $N \circ g = g \circ N$, we call N a **normal subgroup** of G and write $N \triangleleft G$. If $N \triangleleft G$, then let $G|N = $ the set of all cosets (both right and left) of N in G.

EXAMPLES 6.8

1. Consider the group $(Z, +)$ and the subgroup $E \subseteq Z$; E is the subgroup of even integers. The answer to Exercise 1 of the last set shows that E is a normal subgroup of Z, $E \triangleleft Z$, and $Z|E = \{E, 0\}$.

Cosets, normal subgroups, and Lagrange's theorem

2. $\{e\}$ is a subgroup of any group G. For any $g \in G$, the right coset $\{e\} \circ g = \{e \circ g\} = \{g\}$. The left coset $g \circ \{e\} = \{g\}$ also. For any group G, $\{e\} \triangleleft G$ and $G|\{e\} = \{\{g\} | g \in G\}$.

3. G itself is a subgroup of G. We can also show that $G \triangleleft G$ and $G|G = \{G\}$. You are asked for the proof of this in Exercise 2.

4. If G is an abelian group ($x \circ y = y \circ x$ for every x and y in G), then every subgroup of G is normal (Exercise 1).

5. Let G be any group and let $C = \{x \in G | x \circ g = g \circ x \text{ for every } g \in G\}$ be the center of G. We showed that C is a subgroup of G in an exercise in Chapter 5. We will now show that $C \triangleleft G$. For $g \in G$, $C \circ g = \{x \circ g | x \in C\}$. However, for every $x \in C$, $x \circ g = g \circ x$ and therefore

$$C \circ g = \{x \circ g | x \in C\} = \{g \circ x | x \in C\} = g \circ C.$$

Since this is true for any $g \in G$, $C \triangleleft G$.

Note that for a non-abelian group G and a normal subgroup $N \triangleleft G$, the fact that $N \circ g = g \circ N$ does not guarantee that $n \circ g = g \circ n$ for every $n \in N$. Instead, for every $n \circ g \in N \circ g$, there exists an $n' \in N$ so that $n \circ g = g \circ n'$, which is the corresponding element in $g \circ N$.

A very important property of normal subgroups is described in the following theorem.

THEOREM 6.9
Let N be a normal subgroup of G and let g_1 and g_2 be any two elements of G. For any elements $g_1' \in N \circ g_1$ and $g_2' \in N \circ g_2$,

$$(g_1' \circ g_2') \in N \circ (g_1 \circ g_2).$$

Proof: Let $g_1' \in N \circ g_1$ and $g_2' \in N \circ g_2$. Then $g_1' = n_1 \circ g_1$ and $g_2' = n_2 \circ g_2$ for some n_1 and n_2 in N. Now, look at

$$g_1' \circ g_2' = (n_1 \circ g_1) \circ (n_2 \circ g_2)$$
$$= n_1 \circ (g_1 \circ n_2) \circ g_2.$$

Since N is normal, $g_1 \circ N = N \circ g_1$ and therefore $g_1 \circ n_2 = n_2' \circ g_1$ for some $n_2' \in N$. Thus

$$g_1' \circ g_2' = n_1 \circ (g_1 \circ n_2) \circ g_2$$
$$= n_1 \circ (n_2' \circ g_1) \circ g_2$$
$$= (n_1 \circ n_2') \circ (g_1 \circ g_2).$$

Since $n_1 \circ n_2' \in N$, $(g_1' \circ g_2') = (n_1 \circ n_2') \circ (g_1 \circ g_2) \in N \circ (g_1 \circ g_2)$. ∎

This theorem tells us that if N is a normal subgroup, then the product of any element of $N \circ g_1$ and any element of $N \circ g_2$ must be an element of $N \circ (g_1 \circ g_2)$. In fact, having this property also guarantees that a subgroup N is normal. That is, the fact that N is a normal subgroup is equivalent to the fact that the product of elements from $N \circ g_1$ and $N \circ g_2$ is an element of $N \circ (g_1 \circ g_2)$ (Exercise 6). This property allows us to define an operation on the cosets of a normal subgroup N in G:

$$(N \circ g_1) \circ (N \circ g_2) = N \circ (g_1 \circ g_2).$$

Note that there could be a problem with this operation. If $N \circ g_1' = N \circ g_1$ and $N \circ g_2' = N \circ g_2$, then

$$(N \circ g_1) \circ (N \circ g_2) = N \circ (g_1 \circ g_2)$$

while

$$(N \circ g_1') \circ (N \circ g_2') = N \circ (g_1' \circ g_2').$$

Are these "products" equal or does the resultant "product" differ with the representative chosen? That is, is it possible that there is more than one value for $(N \circ g_1) \circ (N \circ g_2)$? Theorems 6.9 and 6.5 provided an answer to that question. By Theorem 6.9 $g_1' \circ g_2' \in N \circ (g_1 \circ g_2)$ and Theorem 6.5 guarantees that $N \circ (g_1' \circ g_2') = N \circ (g_1 \circ g_2)$. The definition of this operation does not depend on the particular representative chosen. This cannot be done if N is not normal; the operation is not well defined in that case.

EXAMPLES 6.9

1. Consider the group $S_3 = \{f_0, f_1, f_2, f_3, f_4, f_5\}$ and the subgroup $H = \{f_0, f_1\}$ with right cosets

$$H \circ f_0 = H \circ f_1 = \{f_0, f_1\}$$
$$H \circ f_2 = H \circ f_3 = \{f_2, f_3\}$$
$$H \circ f_4 = H \circ f_5 = \{f_4, f_5\}.$$

Since $f_2 \circ H = \{f_2, f_4\} \neq H \circ f_2$, H is not normal in S_3. If we try to define a product of two right cosets, as we have for cosets of a normal subgroup, we notice that

$$(H \circ f_2) \circ (H \circ f_4) = H \circ (f_2 \circ f_4) = H \circ f_1$$

while

$$(H \circ f_2) \circ (H \circ f_5) = H \circ (f_2 \circ f_5) = H \circ f_3.$$

Since $H \circ f_1 \neq H \circ f_3$ and the left-hand side of both equations involve the same cosets, the operation is not well defined.

2. However, if in S_3, we chose the subgroup $N = \{f_0, f_3, f_4\}$, we see that
$$N \circ f_0 = N \circ f_3 = N \circ f_4 = \{f_0, f_3, f_4\}$$
$$f_0 \circ N = f_3 \circ N = f_4 \circ N = \{f_0, f_3, f_4\}$$
and
$$N \circ f_1 = N \circ f_2 = N \circ f_5 = \{f_1, f_2, f_5\}$$
$$f_1 \circ N = f_2 \circ N = f_5 \circ N = \{f_1, f_2, f_5\}$$
or that N is normal. $G|N = \{N, N \circ f_1\}$. We can define
$$N \circ N = N$$
$$N \circ (N \circ f_1) = (N \circ f_1) \circ N = N \circ f_1$$
$$(N \circ f_1) \circ (N \circ f_1) = (N \circ f_0) = N.$$

With a normal subgroup, the problems of the previous example do not exist and the operation is well defined.

THEOREM 6.10

If G is a group and $N \triangleleft G$, then the system $(G|N, \circ)$ where $\circ$ is defined by
$$(N \circ g_1) \circ (N \circ g_2) = N \circ (g_1 \circ g_2)$$
is also a group, called the **quotient group.**

Proof: By Theorems 6.9 and 6.5 and the argument above, $(N \circ g_1) \circ (N \circ g_2)$ is well defined and therefore $\circ$ is a binary operation on $G|N$.

Verifying the rest of the group properties is straightforward.
$$(N \circ e) \circ (N \circ g) = (N \circ g) \circ (N \circ e) = N \circ (g \circ e) = N \circ g$$
so that $N \circ e = N$ acts as an identity element.

The associative law follows also, for
$$[(N \circ g_1) \circ (N \circ g_2)] \circ (N \circ g_3) = [N \circ (g_1 \circ g_2)] \circ (N \circ g_3)$$
$$= N \circ [(g_1 \circ g_2) \circ g_3]$$
while
$$(N \circ g_1) \circ [(N \circ g_2) \circ (N \circ g_3)] = (N \circ g_1) \circ [N \circ (g_2 \circ g_3)]$$
$$= N \circ [g_1 \circ (g_2 \circ g_3)]$$
$$= N \circ [(g_1 \circ g_2) \circ g_3]$$

by the associativity of the group operation on G. Since the two expressions are equal, $\circ$ is associative on $G|N$.

Finally $(N \circ g) \circ (N \circ g^{-1}) = N \circ (g \circ g^{-1}) = N \circ e$ and $(N \circ g^{-1}) \circ (N \circ g) = N \circ e$ also so that the inverse element of coset $N \circ g$ is the coset $N \circ g^{-1}$.

Therefore $(G|N, \circ)$ satisfies the definition of a group. ∎

The group $G|N$ is important so we will look at some examples.
EXAMPLES 6.10
1. Let $G = \{e, a, b, c\}$ be the Klein four-group (Example 3.2). Let $N = \{e, a\}$. N is a subgroup of G, and a normal subgroup because G is abelian. There are two distinct cosets

$$e \circ N = N \circ e = \{e, a\} = a \circ N = N \circ a = N$$

and

$$b \circ N = N \circ b = \{b, c\} = c \circ N = N \circ c.$$

The group table for $(G|N, \circ)$ is

$\circ$	N	$N \circ b$
N	N	$N \circ b$
$N \circ b$	$N \circ b$	N

the table for a cyclic group of order 2.

2. For the next example, let Z be the group of integers under addition and let H_5 be the subgroup of all integers divisible by 5. $H_5 = \{\ldots, -5, 0, 5, 10, \ldots\}$. H_5 is a normal subgroup (why?) of Z, so we can look at the cosets and determine the group $Z|H_5$. Example 6.4.2 and Theorem 6.8 tell us what the cosets of H_5 in Z are:

$[0] = H_5 + 0 = H_5 = \{\ldots, -5, 0, 5, 10, \ldots\}$
$[1] = H_5 + 1 = \{\ldots, -4, 1, 6, 11, \ldots\}$
$[2] = H_5 + 2 = \{\ldots, -3, 2, 7, 12, \ldots\}$
$[3] = H_5 + 3 = \{\ldots, -2, 3, 8, 13, \ldots\}$
$[4] = H_5 + 4 = \{\ldots, -1, 4, 9, 14, \ldots\}.$

(Note that the operation in Z is addition.)
Next we construct a group table using the operation

$$(H_5 + a) \circ (H_5 + b) = H_5 + (a + b).$$

$\circ$	H_5	$H_5 + 1$	$H_5 + 2$	$H_5 + 3$	$H_5 + 4$
H_5	H_5	$H_5 + 1$	$H_5 + 2$	$H_5 + 3$	$H_5 + 4$
$H_5 + 1$	$H_5 + 1$	$H_5 + 2$	$H_5 + 3$	$H_5 + 4$	H_5
$H_5 + 2$	$H_5 + 2$	$H_5 + 3$	$H_5 + 4$	H_5	$H_5 + 1$
$H_5 + 3$	$H_5 + 3$	$H_5 + 4$	H_5	$H_5 + 1$	$H_5 + 2$
$H_5 + 4$	$H_5 + 4$	H_5	$H_5 + 1$	$H_5 + 2$	$H_5 + 3$

Cosets, normal subgroups, and Lagrange's theorem

Looking closely, it is readily apparent that this is a table for a cyclic group of order 5. In fact, it is the same as the tables on pp. 20 and 60, for G_5 and the group of equivalence classes of $\equiv_5$, respectively. (Why?)

We leave as an exercise the proof that $Z|H_n$ is a cyclic group of order n, where H_n is the subgroup of all integers exactly divisible by n.

EXERCISES

1. Prove that every subgroup of an abelian group is normal.
2. Show that for any group G:
   a. $G \triangleleft G$ and
   b. $G|G = \{G\}$.
3. Prove that a subgroup N of a group G is normal if and only if $g^{-1} \circ N \circ g = N$ for all g in G.
4. Let N_1 and N_2 be normal subgroups of G. Prove that $N_1 \cap N_2$ is normal in G.
5. Prove the converse of Theorem 6.9.
6. a. Let G be a finite group of order $2n$ and H a subgroup of order n. Prove that H is normal.
   b. If G is a group of order $3n$, show that a subgroup of order n need not be normal.
7. If G is a group, $N \triangleleft G$, and $H < G$, show that $NH = \{n \circ h \mid n \in N$ and $h \in H\}$ is a subgroup of G.
8. In the group $N_{2 \times 2} = \left\{ \begin{pmatrix} a & b \\ c & d \end{pmatrix} \Big| ad - bc \neq 0 \right\}$ under $\cdot$, let $N' = \left\{ \begin{pmatrix} a & b \\ c & d \end{pmatrix} \Big| ad - bc = 1 \right\}$. Show that N' is a normal subgroup of $N_{2 \times 2}$.
9. Show that $Z|H_n$ is a cyclic group of order n.
10. Prove: If G is a group and $N \triangleleft G$ and if H is any subgroup of G with $N \subseteq H$, then $N \triangleleft H$. (The converse of this problem is not true.)
11. Prove that if G is abelian and $N \triangleleft G$, then $G|N$ is abelian.
12. a. If G is a group and $N \triangleleft G$, then if $H < G$, $N \triangleleft H$.
    b. Prove that $H|N$ is a subgroup of $G|N$.

LAGRANGE'S THEOREM

To review, the main results that we have developed for cosets are that:
1. Distinct right cosets are disjoint. That is, right cosets are either identical or they have no elements in common. (This is also true for left cosets.)
2. Every coset of a subgroup H (either right or left) has the same number of elements as H.
3. Every element of G is in exactly one right coset.

Using these observations, we can prove Lagrange's Theorem which gives us some very important information about subgroups and groups.

THEOREM 6.11: LAGRANGE'S THEOREM

Let G be a group of order n and let H be a subgroup of order m; then m divides n.

Proof: Suppose there are exactly k distinct right cosets of H in G (k different nonoverlapping cosets). Choose coset representatives $g_1, g_2, \ldots, g_k$. For any g in G, $H \circ g$ is one of the cosets of the set $\{H \circ g_1, H \circ g_2, \ldots, H \circ g_k\}$. We also know that each element g in G is in exactly one coset. That is,

$$G = (H \circ g_1) \cup (H \circ g_2) \cup \cdots \cup (H \circ g_k).$$

Now each coset has m elements in it and there are k cosets, so that the union contains mk distinct elements. (The elements are distinct because the cosets are disjoint.) But G has order n, so that we have counted the elements in G in two ways. Therefore $|G| = n = mk$. ■

COROLLARY 1

A group G has no subgroups other than $\{e\}$ and G itself if and only if G is cyclic and of prime order.

Proof: Let H be a subgroup of G. If H has order m, then, since m divides p by Lagrange's Theorem and p is prime, m must be equal to 1 or p. If $m = 1$, then $H = \{e\}$, while if $m = p$, $H = G$. Finally, for any $g \in G$, $g \neq e$, $\langle g \rangle$ is a subgroup of G and since $g \in \langle g \rangle$, $\langle g \rangle \neq \{e\}$. Therefore $\langle g \rangle = G$ and G is cyclic.

The converse was presented as a problem in an exercise set in Chapter 5. ■

We make the following definition as a notational convenience.

DEFINITION

If G is a group and $H < G$, then **the index of H in G**, $[G : H]$, is the number of distinct right cosets of H in G.

EXAMPLES 6.11

In S_3, $\{f_0, f_2\}$ is of index 3 while $\{f_0, f_3, f_4\}$ is of index 2.

COROLLARY 2

$$[G : H] = \frac{|G|}{|H|}.$$

The proof is left as an exercise.

Cosets, normal subgroups, and Lagrange's theorem

In the special case where N is a normal subgroup of G, $[G:N] = |G|N|$ and we have the relationship

$|G| = |N| \cdot |G|N|$.

EXERCISES

1. Let $G = \{e, a, b, c\}$ be the Klein four-group. Let $H = \{e, b\}$ be a subgroup of order 2. Show that $G = H \cup (H \circ a)$, and $H \cap (H \circ a) = \emptyset$.
2. Let $G_8 = \{0, 1, 2, 3, 4, 5, 6, 7\}$ be the group of integers modulo 8. Let $H = \{0, 2, 4, 6\}$ be a subgroup of order 4. Show that $G = H \cup (H + 3)$ and $H \cap (H + 3) = \emptyset$.
3. $K = \{0, 4\}$ is a subgroup of G_8 also. Write G_8 as a union of 4 disjoint right cosets.
4. What are the possible orders for subgroups of groups of orders 8, 16, 18, 21, 23, 35, 40?
5. If G is a group with the property that $g^2 = e$ for all $g \in G$, show that G is abelian.
6. Let G be a group of order n. Prove that $g^n = e$ for all $g \in G$. (*Hint:* Let k = order of g. Show that $k|n$.)
7. Show that if a group G is not cyclic, then every g in G generates a proper subgroup.
8. Show that if $H < G$, the number of right cosets of H in G is equal to the number of left cosets of H in G.
9. Verify that $[G:H] = \dfrac{|G|}{|H|}$ for $H < G$.
10. Show that if $K < H < G$, then $[G:H] \cdot [H:K] = [G:K]$.

Using the results we have developed so far, we can essentially classify all groups of order ≤ 7. This will be done in the next exercise set.

By the corollary to Lagrange's Theorem, any group of prime order is cyclic and therefore the only groups of order 2, 3, 5, and 7 are the cyclic groups. What about groups of order 1, 4, and 6? There are certainly cyclic groups of those orders. Are there others?

For example, consider a group G of order 4, with elements e, a, b, and c. If G is not cyclic, then the order of each of its cyclic subgroups, $\langle a \rangle$, $\langle b \rangle$, and $\langle c \rangle$ must be 2. (Why?) Any group table of a noncyclic group of order 4 must have the entries

∘	e	a	b	c
e	e	a	b	c
a	a	e		
b	b		e	
c	c			e

Exercise 5 asks you to find all possible ways to fill in the blank positions in the table so that the resulting system is a group. Exercise 3 indicates an approach to constructing group tables.

EXERCISES

1. Explain a method of determining whether a finite group is abelian by looking at its group table. Prove your assertion.
2. If $G = \langle g \rangle = \{e, g, g^2, \ldots, g^{n-1}\}$ is a finite cyclic group, distinguish a pattern in the group table for G:

$\circ$	e	g	g^2	$\cdots$	g^{n-1}
e					
g					
g^2					
$\vdots$					
g^{n-1}					

3. If G is a finite group of order n, prove that each row and each column of the group table for G contains n distinct elements (a permutation of the elements of G).
4. What must a group table of a group of order 1 look like?
5. Find all possible noncyclic groups of order 4 with elements e, a, b, c. That is, find all possible ways to fill in the table in the text.
6. Consider a noncyclic group $G = \{e, a, b, c, d, f\}$ of order 6 with identity element e.
   a. What are the possible orders of the cyclic subgroups $\langle a \rangle$, $\langle b \rangle$, $\langle c \rangle$, $\langle d \rangle$, $\langle f \rangle$?
   b. Is it possible that all the cyclic subgroups of G except $\langle e \rangle$ have order 3? Prove your result.
   c. Is it possible that all the cyclic subgroups of G except $\langle e \rangle$ have order 2? Prove your result.
   d. Suppose that the order of $\langle a \rangle$ is 2 and the order of $\langle d \rangle$ is 3. Is it possible for a noncyclic group of order 6 to have this situation? Prove your result.
7. Write out a group table for an abelian, but not cyclic group of order 8.
8. Write out a group table of a non-abelian group of order 8.
9. Prove that $N_{2 \times 2} = \left\{ \begin{pmatrix} a & b \\ c & d \end{pmatrix} \middle| ad - bc \neq 0 \right\}$, under the operation of multiplication, is not abelian.
10. Prove that every subgroup of an abelian group is abelian.

Isomorphisms, automorphisms, homomorphisms

In this chapter we will enter a more sophisticated area of group theory. We will be developing some ideas that are more abstract than the ones we have investigated previously. The result will yield deeper information concerning the nature of groups in general and finite groups in particular. We begin by considering mappings from one group into another.

ISOMORPHISMS

Let us look at the group tables for two very similar groups of order 3 and make some observations. Let $G = \{e, g, g_1\}$ and $H = \{f, h, h_1\}$ be groups of order 3 with the tables

$\circ$	e	g	g_1
e	e	g	g_1
g	g	g_1	e
g_1	g_1	e	g

and

$\circ$	f	h	h_1
f	f	h	h_1
h	h	h_1	f
h_1	h_1	f	h

Here e is the identity of G and f is the identity of H. A glance at the tables shows us the similarity between the two groups: if, in the table for G, we replace e by f, g by h, and g_1 by h_1, we get exactly the table for H. In other words, if we change the names of the elements of G, we get the group table for H, or the product in G is replaced by the product in H. This relationship does not always exist as we can see by looking at the following group tables of order 4:

The Klein four-group   and   a cyclic group of order 4

∘	e	a	b	c
e	e	a	b	c
a	a	e	c	b
b	b	c	e	a
c	c	b	a	e

∘	f	x	y	z
f	f	x	y	z
x	x	y	z	f
y	y	z	f	x
z	z	f	x	y

If in the first table we replace e by f, a by x, b by y, and c by z, we do not get the second table for the cyclic group at all, but, instead,

∘	f	x	y	z
f	f	x	y	z
x	x	f	z	y
y	y	z	f	x
z	z	y	x	f

Therefore this cyclic group of order 4 and the Klein four-group have differences other than the names of the elements while the two groups of order 3 do not. This ties in with the insight we gained in the last chapter since the cyclic group of order 4 has an element of order 4 while the elements of the Klein four-group have orders 1 or 2. On the other hand, by the first corollary to Lagrange's Theorem, every group of order 3 is cyclic and therefore all groups of order 3 have the same internal structure.

To formalize this concept so that we can work with it, we make the following definitions.

Let us recall that an **injection** (or 1–1 mapping) is a function with the property that two distinct elements cannot have the same image. It is characterized in two ways: 1. $x \neq y \Rightarrow \alpha(x) \neq \alpha(y)$ or the contra positive, 2. $\alpha(x) = \alpha(y) \Rightarrow x = y$.

A **surjection** (or onto mapping) $\alpha: G \to H$ is a function with the property that every element in the codomain H has a preimage in the domain, G; α maps G onto H (or α is surjective) if for every $h \in H$, there is a $g \in G$ such that $\alpha(g) = h$.

A mapping that is both injective and surjective (1–1 and onto) is called a **bijection** or a 1–1 correspondence. Throughout this text we will use the terms "injection," "surjection," and "bijection" interchangeably with "1–1," "onto," and "1–1 correspondence," respectively.

Isomorphisms, automorphisms, homomorphisms

DEFINITION

Let G and H be any groups and let $\alpha: G \to H$ be a bijection (1–1 and onto mapping) between the elements of G and the elements of H such that for all x, y in G,

$\alpha(x \circ y) = \alpha(x) \circ \alpha(y)$.

Then α is called an **isomorphism** of G onto H and G and H are said to be **isomorphic**.

We re-emphasize here that the definition of an isomorphism always requires that the mapping be bijective (1–1 and onto) and that it preserve the group operation. A note is in order. We used the notation $\circ$ for the composition in both the groups G and H in our defining relation, $\alpha(x \circ y) = \alpha(x) \circ \alpha(y)$; $x \circ y$ is a product of elements in G while $\alpha(x) \circ \alpha(y)$ is a product of elements in H. This could lead to some confusion, but the meaning should be clear in context. If we know a different symbol for the operation on a specific group, we will use that instead.

EXAMPLES 7.1

1. For the groups $G = \{e, g, g_1\}$ and $H = \{f, h, h_1\}$ defined at the beginning of this chapter we define

$\alpha(e) = f$
$\alpha(g) = h$
$\alpha(g_1) = h_1$

All 9 "products" are preserved:

$\alpha(e) \circ \alpha(e) = f \circ f = f = \alpha(e \circ e)$
$\alpha(e) \circ \alpha(g) = f \circ h = h = \alpha(e \circ g)$
$\alpha(g) \circ \alpha(e) = h \circ f = h = \alpha(g \circ e)$
$\alpha(e) \circ \alpha(g_1) = f \circ h_1 = h_1 = \alpha(e \circ g_1)$
$\alpha(g_1) \circ \alpha(e) = h_1 \circ f = h_1 = \alpha(g_1 \circ e)$
$\alpha(g) \circ \alpha(g_1) = h \circ h_1 = f = \alpha(g \circ g_1)$
$\alpha(g_1) \circ \alpha(g) = h_1 \circ h = f = \alpha(g_1 \circ g)$
$\alpha(g) \circ \alpha(g) = h \circ h = h_1 = \alpha(g \circ g)$
$\alpha(g_1) \circ \alpha(g_1) = h_1 \circ h_1 = h = \alpha(g_1 \circ g_1)$.

Since the map is bijective also, α is an isomorphism between the groups G and H.

2. However, for the Klein four-group $\{e, a, b, c\}$ and the cyclic group of order 4 $\{f, x, y, z\}$, if we define

$\alpha(e) = f$
$\alpha(a) = x$
$\alpha(b) = y$
$\alpha(c) = z$

we see that $\alpha(a) \circ \alpha(a) = x \circ x = y$ while $\alpha(a \circ a) = \alpha(e) = f$. Since these "products" are not the same, α cannot be an isomorphism. We leave as an exercise that no bijection between the Klein four-group and a cyclic group of order 4 can be an isomorphism.

Essentially two groups are isomorphic if they are the same except for the notation used.

EXERCISES

1. If $\alpha: G \to H$ is an isomorphism, then $|G| = |H|$.
2. Prove that the Klein four-group is not isomorphic to the cyclic group of order 4. (*Hint:* Show that no bijection preserves the operations.)
3. If G and H are groups of order 2, prove that G is isomorphic to H.
4. Let G be a cyclic group generated by g, $G = \langle g \rangle$. Show that the mapping α, defined by $\alpha(g^n) = g^{-n}$ for any $g^n \in G$, is an isomorphism of G with itself. (*Hint:* Consider the cases of $|G| < \infty$ and $|G| = \infty$ separately.)
5. Find all the isomorphisms of a cyclic group of order 3 with itself.
6. Find all the isomorphisms of the Klein four-group with itself.
7. Let $\alpha: G \to H$ be an isomorphism. Show that α must map the identity of G to the identity of H.
8. If p is a prime, show that any two groups of order p are isomorphic.
9. Let G be a finite abelian group of odd order. Show that the mapping α, defined by $\alpha(x) = x^2$, is an isomorphism of G with itself.
10. If G and H are isomorphic groups, prove that G is abelian if and only if H is abelian.

Before continuing, we pause to prove some useful results about isomorphisms of groups.

THEOREM 7.1

Let G and H be groups and let $\alpha: G \to H$ be an isomorphism.
1. If e is the identity of G, then $\alpha(e)$ is the identity of H.
2. If $x \in G$, then $\alpha(x^{-1}) = \alpha(x)^{-1}$.

Proof of 1: Exercise 7 of the last set.

Isomorphisms, automorphisms, homomorphisms 79

Proof of 2: For $x \in G$, $x \circ x^{-1} = e$. Since α is an isomorphism, $\alpha(x \circ x^{-1}) = \alpha(e) =$ the identity in H.

But

$\alpha(e) = \alpha(x \circ x^{-1})$
$= \alpha(x) \circ \alpha(x^{-1})$

so that $\alpha(x^{-1})$ must be the inverse of $\alpha(x)$ or $\alpha(x)^{-1} = \alpha(x^{-1})$. ∎

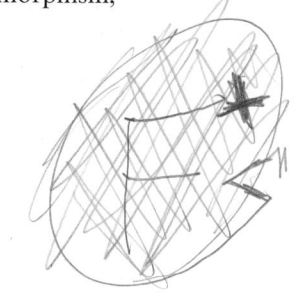

Now let us use the concept of isomorphism to explore the cyclic groups. Recall that under the operation of addition modulo n, we have shown that for every positive integer n there is a cyclic group of order n. The next theorem says that any two cyclic groups of the same order are essentially the same.

THEOREM 7.2
Any two cyclic groups of the same order are isomorphic.

Proof: Let G and H be cyclic groups of the same order.

Case 1: Suppose the orders of $G = \langle g \rangle$ and $H = \langle h \rangle$ are infinite. By Theorem 5.8 we can write

$$G = \{\ldots, g^{-2}, g^{-1}, e = g^0, g, g^2, \ldots\}$$

and

$$H = \{\ldots, h^{-2}, h^{-1}, f = h^0, h^1, h^2, \ldots\}.$$

e and f are the identities of G and H, respectively.

We define a bijection from G to H by

$\alpha(g^k) = h^k \quad$ for every integer k.

We must check that all "products" are preserved.

We need to show that

$\alpha(g^l \circ g^m) = \alpha(g^l) \circ \alpha(g^m)$

for all integers l and m. Now $g^l \circ g^m = g^{l+m}$ so that

$\alpha(g^l \circ g^m) = \alpha(g^{l+m}) = h^{l+m}$

while

$\alpha(g^l) \circ \alpha(g^m) = h^l \circ h^m = h^{l+m}$.

Since the products are the same for all integers l and m, we have shown that the operation is preserved. The proof that α is a bijection is left as an exercise. Therefore any two cyclic groups of infinite order are isomorphic.

Case 2: Assume G and H are cyclic groups of finite order n; $G = \langle g \rangle$ and $H = \langle h \rangle$. By Theorem 5.7 we can write

$$G = \{e, g, g^2, \ldots, g^{n-1}\}$$
$$H = \{f, h, h^2, \ldots, h^{n-1}\}.$$

Again we define a bijection $\alpha(g^k) = h^k$ for $0 \leq k \leq n-1$ and we have to check that products are preserved. For any integers l and m, $g^l \circ g^m = g^{(l+m) \bmod n}$ so that

$$\alpha(g^l \circ g^m) = \alpha(g^{(l+m) \bmod n}) = h^{(l+m) \bmod n}.$$

Since

$$\alpha(g^l) \circ \alpha(g^m) = h^l \circ h^m = h^{(l+m) \bmod n}$$

also, products are preserved and our result is proved. ∎

As a corollary, we have the following result which further condenses the subject of cyclic groups.

COROLLARY
1. Any cyclic group of infinite order is isomorphic to the additive group of integers $(Z, +)$.
2. Any cyclic group of finite order n is isomorphic to the group $(G_n, \oplus)$, $G_n = \{0, 1, \ldots, n-1\}$.

Proof of 1: Let $G = \langle g \rangle$ be a cyclic group with $|G| = \infty$. $(Z, +)$ is a cyclic group of infinite order since $Z = \langle 1 \rangle$. By Theorem 7.2 these two groups are isomorphic.

Proof of 2: If G is a cyclic group of order n, by Theorem 7.2, G is isomorphic to G_n, another cyclic group of order n. ∎

EXERCISES
1. Let G be a cyclic group of infinite order, $G = \langle g \rangle$. What elements of G other than g generate G?
2. Let G be a finite cyclic group of prime order p. Which elements of G serve as generators?
3. If G is a cyclic group of order n, which elements would generate G?
4. For any cyclic groups $G = \langle g \rangle$ and $H = \langle h \rangle$ of the same order, prove that the mapping $\alpha(g^k) = h^k$ for any $g^k \in G$ is a bijection.

AUTOMORPHISMS

DEFINITION

An isomorphism $\alpha: G \to G$ between a group G and itself is called an **automorphism** of G.

EXAMPLES 7.2

1. Let $G = \{e, g, g^2\}$ be the cyclic group of order 3 with group table

$\circ$	e	g	g^2
e	e	g	g^2
g	g	g^2	e
g^2	g^2	e	g

Consider the mapping α, given by

$\alpha(e) = e$
$\alpha(g) = g^2$
$\alpha(g^2) = g.$

To see that α is an isomorphism of G with itself, it is necessary to determine that for any pair of elements in G, α preserves the group operation. There are nine pairs of elements; we see that

$\alpha(e) \circ \alpha(e) = \alpha(e)$
$\alpha(e) \circ \alpha(g) = \alpha(g)$
$\alpha(e) \circ \alpha(g^2) = \alpha(g^2)$
$\alpha(g) \circ \alpha(e) = \alpha(g)$
$\alpha(g) \circ \alpha(g) = \alpha(g^2)$
$\alpha(g) \circ \alpha(g^2) = \alpha(e)$
$\alpha(g^2) \circ \alpha(e) = \alpha(g^2)$
$\alpha(g^2) \circ \alpha(g) = \alpha(e)$
$\alpha(g^2) \circ \alpha(g^2) = \alpha(g)$

so that α is an automorphism of G.

2. Consider any group G and define a mapping $i: G \to G$ by $i(g) = g$ for every g in G. i is called the **identity mapping**. To see that i is an automorphism, we need to verify that (1) i is a bijection (1–1 and onto) and (2) i preserves the operation in G; that is, $i(x \circ y) = i(x) \circ i(y)$. This is left as an exercise.

DEFINITION
If G is a group, let $\mathscr{A}(G)$ be the set of all automorphisms of G.

One of the reasons that we study the automorphisms of a group G is that we can define an operation o on $\mathscr{A}(G)$ that makes the system $(\mathscr{A}(G), \circ)$ a group itself. $\mathscr{A}(G)$ has many interesting properties which we explore both here and in the exercises.

EXAMPLE 7.3
Consider the cyclic group $G = \{e, g, g^2\}$. In the last example we saw that the mapping α, defined by

$\alpha(e) = e$

$\alpha(g) = g^2$

$\alpha(g^2) = g$

is an automorphism of G. The identity map

$i(e) = e$

$i(g) = g$

$i(g^2) = g^2$

is another automorphism of G. We can see that these are the only possibilities since it is necessary to map $e \to e$ (why?) and therefore g must map to either g (as in i) or g^2 (as in α); g^2 must map to the remaining element to make the map bijective. Therefore $\mathscr{A}(G) = \{i, \alpha\}$. As we define a group operation on $\mathscr{A}(G)$, we recall that any group of order 2 must be cyclic and therefore the group table for $(\mathscr{A}(G), \circ)$ must be

$\circ$	i	α
i	i	α
α	α	i

In general, we will define an operation on $\mathscr{A}(G)$ in the following way: If α and β belong to $\mathscr{A}(G)$, we define the map $\beta \circ \alpha$ by

$(\beta \circ \alpha)(g) = \beta(\alpha(g))$

for every $g \in G$. That is, $\beta \circ \alpha$ is the composition of β and α.

THEOREM 7.3
If $\beta, \alpha \in \mathscr{A}(G)$, then $\beta \circ \alpha \in \mathscr{A}(G)$. That is, $\circ$ is a binary operation on $\mathscr{A}(G)$.
  Proof: We need to show that $\beta \circ \alpha$ is a bijection from G to G and that $\beta \circ \alpha$ preserves the operation on G.

Isomorphisms, automorphisms, homomorphisms

Note first that our map $\beta \circ \alpha$ associates with each element $g \in G$ another element $\beta(\alpha(g)) \in G$. Since $g \in G$ and $\alpha: G \to G$, then $\alpha(g) \in G$. Also, since $\beta: G \to G$, $\beta[\alpha(g)] \in G$.

To show that $\beta \circ \alpha$ is injective, consider the following: Let g, h be any elements of G. Then if $\beta(\alpha(g)) = \beta(\alpha(h))$, since β is injective (1–1), we know that $\alpha(g) = \alpha(h)$. Then, since α is injective, we have $g = h$. Thus, if $(\beta \circ \alpha)(g) = (\beta \circ \alpha)(h)$, we must have $g = h$ and $\beta \circ \alpha$ is injective.

To see that $\beta \circ \alpha$ is surjective (onto G), consider any $g \in G$. We must be able to find $h \in G$ such that $(\beta \circ \alpha)(h) = g$. But since β is surjective, there exists $g' \in G$ such that $\beta(g') = g$, and finally since α is surjective, there is an $h \in G$ such that $\alpha(h) = g'$. Thus $(\beta \circ \alpha)(h) = \beta(g') = g$. Since this is true for any $g' \in G$ we have that $\beta \circ \alpha$ is surjective.

The final step in proving that $\beta \circ \alpha$ is an automorphism is to verify that the operation in G is preserved by the composition of automorphisms. That is, if $g, h \in G$, and α and β are automorphisms of G, is

$$(\beta \circ \alpha)(g \circ h) = (\beta \circ \alpha)(g) \circ (\beta \circ \alpha)(h)?$$

If so, then $\beta \circ \alpha$ will be an automorphism whenever α and β are.

Since α and β are automorphisms, we can write

$$\begin{aligned}
(\beta \circ \alpha)(g \circ h) &= \beta(\alpha(g \circ h)) \\
&= \beta(\alpha(g) \circ \alpha(h)) \quad \text{(since } \alpha \text{ is an automorphism)} \\
&= \beta(\alpha(g)) \circ \beta(\alpha(h)) \quad \text{(since } \beta \text{ is an automorphism)} \\
&= (\beta \circ \alpha)(g) \circ (\beta \circ \alpha)(h)
\end{aligned}$$

so that $\beta \circ \alpha$ is indeed an automorphism. ∎

We will now proceed to show that the set of all automorphisms of a group G is itself a group under the operation of composition as defined above.

THEOREM 7.4

$(\mathscr{A}(G), \circ)$ is a group.

Proof: The previous theorem showed that $\circ$ is a binary operation on $(\mathscr{A}(G), \circ)$.

To complete the proof that $(\mathscr{A}(G), \circ)$ is a group, we have to check the defining conditions for a group: associativity, existence of an identity, existence of inverses.

1. Associativity: If $\alpha, \beta, \gamma \in \mathscr{A}(G)$, does $(\alpha \circ \beta) \circ \gamma = \alpha \circ (\beta \circ \gamma)$? Since $(\alpha \circ \beta) \circ \gamma$ and $\alpha \circ (\beta \circ \gamma)$ are mappings from G to G, they are equal if they

have the same effect on every element of G. That is, if $[(\alpha \circ \beta) \circ \gamma](g) = [\alpha \circ (\beta \circ \gamma)](g)$ for every g in G. But

$$[(\alpha \circ \beta) \circ \gamma](g) = (\alpha \circ \beta)(\gamma(g))$$
$$= \alpha[\beta(\gamma(g))] \quad \text{(Why?)}$$

while

$$[\alpha \circ (\beta \circ \gamma)](g) = \alpha[(\beta \circ \gamma)(g)]$$
$$= \alpha[\beta(\gamma(g))] \quad \text{(Why?)}.$$

Thus the two mappings of G to G, $\alpha \circ (\beta \circ \gamma)$ and $(\alpha \circ \beta) \circ \gamma$, have the same effect on every $g \in G$, so they are the same element of $\mathscr{A}(G)$ and the associative law holds.

2. *Existence of an Identity Element*: The identity element of $\mathscr{A}(G)$ is the identity map i defined by $i(g) = g$ for all $g \in G$. (Exercise 1 proves that i is an automorphism of G.)

3. *Existence of Inverses*: If $\alpha \in \mathscr{A}(G)$, we define $\alpha^{-1}: G \to G$ by $\alpha^{-1}(g) = h$ if and only if $\alpha(h) = g$. We have to show that $\alpha^{-1} \in \mathscr{A}(G)$ (injective, surjective, and preserves operations) and that $\alpha \circ \alpha^{-1} = \alpha^{-1} \circ \alpha = i$. First, α^{-1} is a mapping of G to G, for, given any $g \in G$, there is a unique $h \in G$ such that $\alpha(h) = g$ (because α is surjective). Thus any g has one image under α^{-1}, namely that h such that $\alpha(h) = g$. Therefore α^{-1} is a well-defined map, $\alpha^{-1}: G \to G$.

Next we show that α is injective. If $\alpha^{-1}(g) = \alpha^{-1}(g_1) = h$, that means $\alpha(h) = g$ and $\alpha(h) = g_1$. However, α is a mapping and $\alpha(h)$ is unique, so $g = g_1$.

Third, α^{-1} is surjective for, given any $h \in G$, $\alpha(h) = g$ for some $g \in G$. Therefore $\alpha^{-1}(g) = h$.

Finally, to see that $\alpha^{-1} \in \mathscr{A}(G)$, we have to show that the operation is preserved; that is, $\alpha^{-1}(g \circ h) = \alpha^{-1}(g) \circ \alpha^{-1}(h)$ for all $g, h \in G$. But for any element x of G, it is true that $\alpha(\alpha^{-1}(x)) = x$. For $\alpha^{-1}(x) = y$ means $\alpha(y) = x$ so $\alpha(\alpha^{-1}(x)) = \alpha(y) = x$. Now look at $\alpha(\alpha^{-1}(g) \circ \alpha^{-1}(h))$. Since α is an automorphism and preserves products,

$$\alpha(\alpha^{-1}(g) \circ \alpha^{-1}(h)) = \alpha(\alpha^{-1}(g)) \circ \alpha(\alpha^{-1}(h))$$
$$= g \circ h.$$

Also, $\alpha(\alpha^{-1}(g \circ h)) = g \circ h$. Thus $\alpha(\alpha^{-1}(g) \circ \alpha^{-1}(h)) = \alpha(\alpha^{-1}(g \circ h))$. Since α is injective (1–1), this implies that

$$\alpha^{-1}(g \circ h) = \alpha^{-1}(g) \circ \alpha^{-1}(h)$$

proving that $\alpha^{-1} \in \mathscr{A}(G)$.

Isomorphisms, automorphisms, homomorphisms

As our last step, we must show that $\alpha^{-1} \circ \alpha = i = \alpha \circ \alpha^{-1}$. We have just seen above that $\alpha \circ \alpha^{-1} = i$. The proof that $\alpha^{-1} \circ \alpha = i$ is similar: for any $x \in G$

$$(\alpha^{-1} \circ \alpha)(x) = \alpha^{-1}(\alpha(x))$$
$$= \alpha^{-1}(y) \quad \text{(for some } y \in G \text{ with } \alpha(x) = y)$$
$$= x \quad \text{(since } \alpha(x) = y \Leftrightarrow \alpha^{-1}(y) = x)$$
$$= i(x).$$

Thus α^{-1} is the inverse of α in $\mathscr{A}(G)$ and we have verified all the conditions for $\mathscr{A}(G)$ to be a group. ■

The properties of automorphism groups are very interesting and we shall explore some of them in the exercises. One of the more interesting properties, which we state and partially prove, is that if G and H are isomorphic, then $\mathscr{A}(G)$ and $\mathscr{A}(H)$ must also be isomorphic. It is interesting to note that the converse is not necessarily true: *nonisomorphic groups can have isomorphic automorphism groups.*

THEOREM 7.5

If G and H are isomorphic, then $\mathscr{A}(G)$ and $\mathscr{A}(H)$ must be isomorphic.

Proof: Let $\psi: G \to H$ be an isomorphism from G onto H. For any $\alpha \in \mathscr{A}(G)$, define α' by

$$\alpha'(h) = \psi[\alpha(\psi^{-1}(h))]$$

or

$$\alpha' = \psi \circ \alpha \circ \psi^{-1}.$$

The proof now has several parts:
1. Show that ψ^{-1} is an isomorphism from H to G.
2. Show that $\alpha' \in \mathscr{A}(H)$.
3. Show that the mapping $\alpha \to \alpha'$ is an isomorphism of $\mathscr{A}(G)$ and $\mathscr{A}(H)$. The proof is fairly computational and contains many of the elements of the proof of Theorem 7.4. We leave its completion as an exercise.

EXERCISES
1. Show that the identity map $i: G \to G$ defined by $i(g) = g$ for every $g \in G$ is an automorphism of any group G.
2. Let $\{e, a, b, c\}$ be the Klein four-group and let β be defined by $\beta(e) = e$, $\beta(a) = b$, $\beta(b) = c$, $\beta(c) = a$. Prove that β is an automorphism of the Klein four-group.
3. Find the automorphism group of the Klein four-group.

4. Find the automorphism groups of the cyclic groups of orders 4, 5, and 6.
5. Prove that an automorphism of a cyclic group is completely determined by its action on a generator.
6. If G is a cyclic group of order p, what is the automorphism group of G?
7. What is the automorphism group of the cyclic group G_n for any integer n?
8. Complete the proof of Theorem 7.5.
9. Prove that if G is isomorphic to H and H is isomorphic to K, then G is isomorphic to K.
10. Find an example of nonisomorphic groups that have isomorphic automorphism groups.
11. Let G be a group and let x be any element of G. Define a function $\alpha_x : G \to G$ by $\alpha_x(g) = x^{-1} \circ g \circ x$.
    a. Prove that $\alpha_x \in \mathscr{A}(G)$ for every $x \in G$. α_x is called the **inner automorphism** of G corresponding to x.
    b. Prove that $\{\alpha_x \mid x \in G\} = \mathscr{I}(G)$ is a subgroup of $\mathscr{A}(G)$.
    c. Show that if G is abelian, then $\mathscr{I}(G) = \{i\}$.
    d. In Chapter 6 we showed that $C = \{x \in G \mid g \circ x = x \circ g \text{ for all } g \in G\}$ is a normal subgroup of G. Prove that G/C is isomorphic with $\mathscr{I}(G)$.
12. Show that the mapping $\alpha : G \to G$, defined by $\alpha(g) = g^{-1}$ for every $g \in G$, is an automorphism if and only if G is abelian.

HOMOMORPHISMS

The concept of homomorphism is a generalization of that of isomorphism. We do not require that a homomorphism be bijective—only that the group operations are preserved.

EXAMPLES 7.4

1. Let $G = \{e, a, b, c\}$ be the Klein four-group and let $H = \{e, g\}$ be a group of order 2. Define a mapping $\alpha : G \to H$ by

$\alpha(e) = e$

$\alpha(a) = e$

$\alpha(b) = g$

$\alpha(c) = g$.

α cannot be an isomorphism because it is not injective (1–1). However, it does preserve the operations:

$\alpha(a \circ b) = \alpha(c) = g$ while $\alpha(a) \circ \alpha(b) = e \circ g = g$ also;

$\alpha(e \circ b) = \alpha(b) = g$ while $\alpha(e) \circ \alpha(b) = e \circ g = g$ also;

$\alpha(b \circ c) = \alpha(a) = e$ while $\alpha(b) \circ \alpha(c) = g \circ g = e$ also;

etc.

All 16 products should be computed to verify this property.

Isomorphisms, automorphisms, homomorphisms

2. As another example, let Z be the additive group of integers and let $H = \{e, g\}$ be the cyclic group of order 2. Define

$$\beta(n) = \begin{cases} e, & \text{if } n \text{ is even} \\ g, & \text{if } n \text{ is odd} \end{cases} \quad \text{for every } n \in Z.$$

Again β cannot be an isomorphism but it does preserve the group operations: $\beta(n + m) = \beta(n) \circ \beta(m)$.

To see this we consider two cases:

1. If n and m are both even or both odd, then $n + m$ is even and

$$\beta(n + m) = e$$

whereas

$$\beta(n) \circ \beta(m) = \begin{cases} e \circ e = e, & \text{if } n \text{ and } m \text{ are both even} \\ g \circ g = e, & \text{if } n \text{ and } m \text{ are both odd.} \end{cases}$$

2. On the other hand, if n is even and m is odd, then $n + m$ is odd so

$$\beta(n + m) = g$$

while

$$\beta(n) \circ \beta(m) = e \circ g = g.$$

The case of n odd and m even is similar. In all cases

$$\beta(n + m) = \beta(n) \circ \beta(m).$$

In both of the above examples, we have defined a function from one group to another which preserves group composition.

DEFINITION

If G and H are groups, a mapping $\alpha: G \to H$ is called a **homomorphism** from G to H if

$$\alpha(g_1 \circ g_2) = \alpha(g_1) \circ \alpha(g_2)$$

for every pair $g_1, g_2 \in G$.

Homomorphisms have many of the same properties that isomorphisms have and are basic to many other concepts. Notice that isomorphisms are a special class of such maps with the additional requirement that they be bijective (1–1 and onto). That is, every isomorphism is a homomorphism, but there are homomorphisms that are not isomorphisms. All the results that we develop will be valid for isomorphisms as well as homomorphisms. We state a theorem that is analogous to Theorem 7.1 for isomorphisms. The proof is exactly the same as the one presented on p. 79 and will not be repeated here.

THEOREM 7.6
Let α be a homomorphism from G to H. Then:
1. $\alpha(e)$ is the identity of H if e is the identity of G.
2. For any $x \in G$, $\alpha(x^{-1}) = \alpha(x)^{-1}$.

Since a homomorphism $\alpha: G \to H$ is not required to be surjective (onto), we are not guaranteed that for every $h \in H$, there is a $g \in G$ such that $\alpha(g) = h$. Consequently, we make the following definition.

DEFINITION
If G and H are groups and $\alpha: G \to H$ is a homomorphism, then $\text{Im}(\alpha) = \{\alpha(g) | g \in G\}$. $\text{Im}(\alpha)$ is read "the **image** of α."

EXAMPLES 7.5
1. Let $G = \{e, g, g^2\}$ be a cyclic group of order 3 and $S_3 = \{f_0, f_1, f_2, f_3, f_4, f_5\}$ be the symmetric group on three elements. Define $\alpha: G \to S_3$ by
$$\alpha(e) = f_0$$
$$\alpha(g) = f_3$$
$$\alpha(g^2) = f_4.$$
We can show that α is a homomorphism by showing that all nine products of pairs of elements of G are preserved by α. $\text{Im}(\alpha) = \{\alpha(g) | g \in G\} = \{f_0, f_3, f_4\}$.

2. If G and H are any groups, we define $\beta: G \to H$ by $\beta(g) = e$, the identity in H for every $g \in G$. β is a homomorphism since for $g, g' \in G$,
$$\alpha(g \circ g') = e = e \circ e = \alpha(g) \circ \alpha(g').$$

Notice that in both examples, $\text{Im}(\alpha)$ is a subgroup of the codomain. The next theorem states that this is always the case.

THEOREM 7.7
If $\alpha: G \to H$ is a homomorphism, then $\text{Im}(\alpha)$ is a subgroup of H.

Proof: Note that $\text{Im}(\alpha) \neq \emptyset$ since the image of the identity of G is in $\text{Im}(\alpha)$. By Theorem 5.2, it is sufficient to show that for any $x, y \in \text{Im}(\alpha)$, $x \circ y^{-1} \in \text{Im}(\alpha)$. To do this, choose any $x, y \in \text{Im}(\alpha)$. Since $x \in \text{Im}(\alpha)$, there must be an element $g \in G$ such that $\alpha(g) = x$. Similarly, we can find $h \in G$ so that $\alpha(h) = y$. Consider the element $\alpha(g \circ h^{-1}) \in \text{Im}(\alpha)$. Since α is a

Isomorphisms, automorphisms, homomorphisms 89

homomorphism,

$$\alpha(g \circ h^{-1}) = \alpha(g) \circ \alpha(h^{-1})$$
$$= \alpha(g) \circ \alpha(h)^{-1} \quad \text{(by Theorem 7.6)}$$
$$= x \circ y^{-1}.$$

Therefore $x \circ y^{-1} \in \text{Im}(\alpha)$ and $\text{Im}(\alpha)$ is a subgroup of H. ∎

Next we investigate a subset of the domain that is important in the study of homomorphisms.

DEFINITION

If $\alpha: G \to H$ is a homomorphism, then

$$\ker(\alpha) = \{g \in G \mid \alpha(g) = e', \text{ the identity of } H\}$$

$\ker(\alpha)$ is called the **kernel** of α.

Note that $\ker(\alpha)$ is never empty since, by Theorem 7.6, the identity e of G maps to the identity of H and therefore $e \in \ker(\alpha)$. Since α is not necessarily injective there may be elements other than e in $\ker(\alpha)$. We will find the kernel of each homomorphism in our previous examples.

EXAMPLES 7.6

1. In the example where α mapped the Klein four-group $\{e, a, b, c\}$ to the cyclic group $\{e, g\}$ by $\alpha(e) = e$, $\alpha(a) = e$, $\alpha(b) = g$, $\alpha(c) = g$, the kernel of the map

$$\ker(\alpha) = \{e, a\}.$$

2. In the homomorphism $\beta: Z \to \{e, g\}$ defined by

$$\beta(n) = \begin{cases} e, & \text{if } n \text{ is even} \\ g, & \text{if } n \text{ is odd} \end{cases}, \quad \ker(\beta) = \{2n \mid n \in Z\} = E$$

the subgroup of Z consisting of the even integers.

3. For the map $\alpha: \{e, g, g^2\} \to S_3$ defined by $\alpha(e) = f_0$, $\alpha(g) = f_3$, $\alpha(g^2) = f_4$, the kernel is $\{e\}$.

4. If, for any group G, $\beta: G \to H$ maps every $g \in G$ to the identity in the group H, then $\ker(\beta) = G$.

5. If $\gamma: G \to H$ is an isomorphism, then Theorem 7.1 guarantees that the identity of G maps to the identity of H.

The injective quality of the isomorphism γ guarantees that the identity of G is the *only* element that maps to the identity of H. Therefore, for any isomorphism γ, $\ker(\gamma) = \{e\}$.

Notice that all of these sets are subgroups of the domain of the homomorphism. Again, this situation is true for any homomorphism.

THEOREM 7.8

Let $\alpha: G \to H$ be a homomorphism with $\ker(\alpha) = \{g \in G \mid \alpha(g) = e\}$. Then $\ker(\alpha)$ is a subgroup of G.

Proof: Suppose that e' is the identity in H. $\ker(\alpha) \neq \emptyset$ since the identity of G belongs to $\ker(\alpha)$. Again by Theorem 5.2 we have to show that for any $g_1, g_2 \in \ker(\alpha)$, $g_1 \circ g_2^{-1} \in \ker(\alpha)$. Choose g_1 and g_2 in $\ker(\alpha)$. Then $\alpha(g_1) = e'$ and $\alpha(g_2) = e'$. Therefore

$$\begin{aligned}\alpha(g_1 \circ g_2^{-1}) &= \alpha(g_1) \circ \alpha(g_2^{-1}) &&\text{(since } \alpha \text{ is a homomorphism)} \\ &= \alpha(g_1) \circ \alpha(g_2)^{-1} &&\text{(by Theorem 7.6)} \\ &= e' \circ e' \\ &= e'.\end{aligned}$$

Since $g_1 \circ g_2^{-1}$ maps to the identity e' of H, $g_1 \circ g_2^{-1} \in \ker(\alpha)$ and $\ker(\alpha)$ is a subgroup of G. ∎

EXERCISES

1. Let α_n be a mapping from the integers $(Z, +)$ to the cyclic group $G_n = \{0, 1, 2, \ldots, n-1\}$ under addition modulo n. Define $\alpha_n(k) = k \bmod n$. Show that α_n is a surjective homomorphism. What is $\ker(\alpha_n)$?

2. If $N_{2 \times 2} = \left\{ \begin{pmatrix} a & b \\ c & d \end{pmatrix} \middle| ad - bc \neq 0 \right\}$ under multiplication, define a mapping $\alpha: N_{2 \times 2} \to \mathbb{R}$ (the reals) by $\alpha(M) = \det(M)$.
   a. Show that α is a homomorphism from $N_{2 \times 2}$ to the group of nonzero reals under multiplication.
   b. What is $\text{Im}(\alpha)$?
   c. What is $\ker(\alpha)$?

3. If $\alpha: G \to H$ is a homomorphism, show that $\ker(\alpha) = \{e\}$ if and only if α is injective.

4. Let $\alpha: G \to H$ be a homomorphism with $\ker(\alpha) = \{e\}$. Prove that G and $\text{Im}(\alpha)$ are isomorphic groups.

5. Let $\alpha: G \to H$ be a homomorphism. Prove:
   a. If G is cyclic, then $\text{Im}(\alpha)$ is cyclic also.
   b. If G is abelian, then $\text{Im}(\alpha)$ is abelian also.
   Next we will investigate the cosets of $\ker(\alpha)$ in G.

THEOREM 7.9

Let G and H be groups and let $\alpha: G \to H$ be a homomorphism. Then $\ker(\alpha)$ is a normal subgroup of G.

Proof: By definition, a normal subgroup N has all its right cosets, $N \circ g$, equal to its left cosets, $g \circ N$, or, equivalently, that $g^{-1} \circ N \circ g = N$

Isomorphisms, automorphisms, homomorphisms 91

for all $g \in G$. (This was the result of Exercise 3 in the subsection on normal subgroups.)

We will use this second formulation to prove that $\ker(\alpha)$ is normal. Let e' be the identity of H. Choose any element $g \in G$.

1. To show that $g^{-1} \circ \ker(\alpha) \circ g \subseteq \ker(\alpha)$, we choose an element $x \in \ker(\alpha)$ and show that $(g^{-1} \circ x \circ g) \in \ker(\alpha)$; that is, we must show that $g^{-1} \circ x \circ g$ maps to the identity e' of H.

$$\begin{aligned}\alpha(g^{-1} \circ x \circ g) &= \alpha(g^{-1}) \circ \alpha(x) \circ \alpha(g) && \text{(since } \alpha \text{ is a homomorphism)} \\ &= \alpha(g^{-1}) \circ e' \circ \alpha(g) && \text{(since } x \in \ker(\alpha)) \\ &= \alpha(g^{-1}) \circ \alpha(g) && \text{(by property of } e') \\ &= e' && \text{(since } \alpha(g^{-1}) = \alpha(g)^{-1}).\end{aligned}$$

Therefore $(g^{-1} \circ x \circ g) \in \ker(\alpha)$ or $g^{-1} \circ \ker(\alpha) \circ g \subseteq \ker(\alpha)$.

2. To show that $\ker(\alpha) \subseteq g^{-1} \circ \ker(\alpha) \circ g$, choose $x \in \ker(\alpha)$. We must show that for some $y \in \ker(\alpha)$, $x = g^{-1} \circ y \circ g$. This is equivalent to showing that for some $y \in \ker(\alpha)$, $g \circ x \circ g^{-1} = y$ or $g \circ x \circ g^{-1} \in \ker(\alpha)$. Again we consider

$$\begin{aligned}\alpha(g \circ x \circ g^{-1}) &= \alpha(g) \circ \alpha(x) \circ \alpha(g^{-1}) && \text{(why?)} \\ &= \alpha(g) \circ e' \circ \alpha(g^{-1}) && \text{(why?)} \\ &= \alpha(g) \circ \alpha(g^{-1}) && \text{(why?)} \\ &= e' && \text{(why?).}\end{aligned}$$

Now since $\ker(\alpha) \subseteq g^{-1} \circ \ker(\alpha) \circ g$ and $g^{-1} \circ \ker(\alpha) \circ g \subseteq \ker(\alpha)$, we have equality of the two sets:

$\ker(\alpha) = g^{-1} \circ \ker(\alpha) \circ g$

or that $\ker(\alpha)$ is a normal subgroup of G. ■

Recall that for any normal subgroup N of a group G, we defined $G|N$ to be the set of all cosets of N:

$G|N = \{N \circ g | g \in G\}$.

With the operation

$(N \circ g) \circ (N \circ h) = N \circ (g \circ h)$

the system $(G|N, \circ)$ is a group.

Since, for any homomorphism $\alpha: G \to H$, $\ker(\alpha)$ is a normal subgroup of G, then $G|\ker(\alpha)$ has a group structure under the operation $\circ$ defined above. The relationships among the various groups are explored in the following theorems.

THEOREM 7.10: HOMOMORPHISM THEOREM FOR GROUPS

Let $\alpha: G \to H$ be a homomorphism; then $G|\ker(\alpha)$ is isomorphic to $\text{Im}(\alpha)$.

Proof: Let $N = \ker(\alpha)$ and consider the mapping ψ from $G|N$ to $\text{Im}(\alpha)$, defined by $\psi(N \circ g) = \alpha(g)$. We want to show that ψ is an isomorphism. We have to prove that ψ is (1) well defined. (2) injective, (3) surjective, and (4) a homomorphism.

1. We can show ψ is well defined by showing that the image of a coset does not depend on the representative chosen. $\psi(N \circ g) = \alpha(g)$; if we choose another representative $g_1 \in N \circ g$, we need to show that $\psi(N \circ g_1) = \psi(N \circ g)$ or, equivalently, $\alpha(g_1) = \alpha(g)$. If this is true, any other coset representative of $N \circ g$ will give the same image under our definition of ψ. Let $g_1 \in N \circ g$. Then $g_1 = n \circ g$ for some $n \in N$; $\alpha(g_1) = \alpha(n \circ g) = \alpha(n) \circ \alpha(g) = e' \circ \alpha(g) = \alpha(g)$ where e' is the identity of H. Therefore α is well defined.

2. ψ is injective means that if $\psi(N \circ g) = \psi(N \circ g')$, then $N \circ g = N \circ g'$. We begin by assuming that $\psi(N \circ g) = \psi(N \circ g')$ which implies that $\alpha(g) = \alpha(g')$. Choosing the element $n = g \circ (g')^{-1} \in G$, we can write $g = n \circ g'$. Then

$\alpha(g) = \alpha(n \circ g')$

$\quad\,\, = \alpha(n) \circ \alpha(g') \qquad$ (since α is a homomorphism).

Also

$\alpha(g) = \alpha(g') \qquad$ (by hypothesis)

$\quad\,\, = e' \circ \alpha(g') \qquad$ (for the identity $e' \in H$).

Therefore the cancellation laws guarantee that $\alpha(n) = e'$ or that $n \in N$, the kernel of α. Therefore $g = n \circ g' \in N \circ g'$ and $N \circ g = N \circ g'$ which is the result we needed.

3. Next, ψ is clearly surjective (onto $\text{Im}(\alpha)$) for if h is any element of $\text{Im}(\alpha)$, then there exists a $g \in G$ such that $\alpha(g) = h$. Then $\psi(N \circ g) = \alpha(g) = h$ so that ψ is surjective.

4. Finally, to prove that ψ is a homomorphism, we have to show that $\psi[(N \circ g) \circ (N \circ g')] = \psi(N \circ g) \circ \psi(N \circ g')$ for all $g, g' \in G$. Now

$\psi[(N \circ g) \circ (N \circ g')] = \psi(N \circ (g \circ g')) = \alpha(g \circ g')$

while

$\psi(N \circ g) \circ \psi(N \circ g') = \alpha(g) \circ \alpha(g') = \alpha(g \circ g')$

since α is a homomorphism. We can conclude that ψ is an isomorphism. ∎

THEOREM 7.11

If G is a group and N is any normal subgroup of G, then the mapping $\alpha: G \to G|N$, defined by

$$\alpha(g) = N \circ g$$

is a surjective homomorphism of G onto $G|N$ with $N = \ker(\alpha)$.

Proof: We need to show that the mapping α is (1) a homomorphism, (2) surjective, and (3) has kernel N.

1. For any elements $g, g_1 \in G$,

$$\alpha(g \circ g_1) = N \circ (g \circ g_1)$$

while

$$\alpha(g) \circ \alpha(g_1) = (N \circ g) \circ (N \circ g_1)$$
$$= N \circ (g \circ g_1)$$

so that α preserves the operations involved and is a homomorphism.

2. For any $N \circ g \in G|N$, $\alpha(g) = N \circ g$ so that $G|N \subseteq \operatorname{Im}(\alpha)$ and α is surjective. ($\operatorname{Im}(\alpha) \subseteq G|N$ by definition.)

3. What is $\ker(\alpha)$? N is the identity of $G|N$ so that if $g \in \ker(\alpha)$, then $\alpha(g) = N$. Also, $\alpha(g) = N \circ g$ and we see that $N = N \circ g$. Therefore $g \in \ker(\alpha)$ is equivalent to $g \in N$, or $\ker(\alpha) \subseteq N$. For $n \in N$, $\alpha(n) = N \circ n = N$ so that $n \in \ker(\alpha)$ or $N \subseteq \ker(\alpha)$. Since we have set containment in both directions, $N = \ker(\alpha)$. ∎

The last two theorems taken together show the connection between the notions of homomorphism and quotient groups. The first says that the image of a homomorphism is essentially the same as a quotient group. The second says that for every normal subgroup, the mapping $g \to N \circ g$ gives a homomorphism with kernel N and image $G|N$.

With the help of the material on homomorphisms, we can prove an additional important property of normal subgroups which we will use in the next chapter. Before we state and prove the theorem, let us review a few properties of normal subgroups.

If G is a group and N is a normal subgroup of G, we write $N \triangleleft G$. Notice that for any other subgroup $H < G$, with $N \subseteq H$, that $N \triangleleft H$ also. Therefore we can consider the quotient group $H|N$ which we found to be a subgroup of $G|N$ (Exercise 13b of the exercise set dealing with normal subgroups). Therefore, for every subgroup of G which contains N, we get a subgroup of $G|N$. The next theorem states that this correspondence is a bijection.

THEOREM 7.12

Let G be a group and let N be a normal subgroup of G. Then there exists a bijection (1–1 correspondence) between the subgroups of $G|N$ and the subgroups of G which contain N.

Proof: If G is a group and $N \triangleleft G$, consider the homomorphism $\alpha: G \to G|N$ defined by $\alpha(g) = N \circ g$. Consider a subgroup $H|N < G|N$ and define the set H by

$$H = \{h \,|\, \alpha(h) \in H|N\}.$$

That is, H is the set of elements of G which map to $H|N$ under the map α. We need to show that (1) H is a subgroup of G which contains N and (2) the mapping ψ defined by $\psi(H|N) = H$ takes the set of all subgroups of $G|N$ bijectively onto the set of all subgroups of G which contain N.

1. Choose x and y in H. Then $\alpha(x \circ y^{-1}) = \alpha(x) \circ \alpha(y)^{-1}$ (why?) and $\alpha(x) \circ \alpha(y)^{-1} \in H|N$. (Why?) Therefore $x \circ y^{-1} \in H$. By Theorem 5.2, H is a subgroup of G. (Why is H nonempty?)

To show that $N \subseteq H$, choose any $n \in N$, then $\alpha(n) = N \in H|N$; it is the identity. By our definition, $n \in H$ and therefore $N \subseteq H$.

2. We need to show that ψ is injective and surjective.

To see that ψ is injective, consider two subgroups $H_1|N$ and $H_2|N$ of $G|N$. If $H_1|N \neq H_2|N$, we will prove that $H_1 \neq H_2$. Since $H_1|N \neq H_2|N$, assume that there is a coset $N \circ g_0$ that belongs to $H_1|N$ and not to $H_2|N$. $N \circ g_0 \in H_1|N$ implies that $g_0 \in H_1$; $N \circ g_0 \notin H_2|N$ implies that $g_0 \notin H_2$. Therefore $H_1 \neq H_2$, which is what we needed to prove.

To show that ψ is surjective, we need to show that every subgroup of G which contains N is the image of some subgroup of $G|N$. Let H be any subgroup of G which contains N. Consider $H|N$. This is a subgroup of $G|N$ and $\psi(H|N) = H$. Therefore ψ is surjective and we have proved the theorem. ∎

EXERCISES

1. Assume that G is a group with normal subgroups N_1 and N_2. If $N_1 < N_2$, prove that $N_2|N_1 \triangleleft G|N_1$.
2. If $\alpha: G \to H$ is a homomorphism and if $K < H$, prove that $\{g \in G \,|\, \alpha(g) \in K\}$ is a subgroup of G.
3. If $\alpha: G \to H$ is a homomorphism and $G = \langle g \rangle$ is a cyclic group, prove that $\mathrm{Im}(\alpha)$ is cyclic also.
4. If Z is the additive group of integers and $Z_n = \{0, 1, 2, \ldots, n-1\}$ is the cyclic group of integers modulo n, prove that $Z/\langle n \rangle$ is isomorphic to Z_n.
5. If $\alpha: G \to H$ is a homomorphism and if G is a finite group, prove $|\mathrm{Im}(\alpha)|$ divides $|G|$.

The structure of finite groups

In this chapter we will deal with the following type of question: If G is a group of order n and if $m|n$, does G have a subgroup of order m? Lagrange's Theorem tells us that if $H < G$, then $|H|\,|\,|G|$, but so far we have no results about the converse situation.

We will first present some results for finite abelian groups and then develop some concepts that help to attack the non-abelian cases. We will conclude with a statement of Sylow's Theorem. The proofs of Cauchy's Theorem for Abelian Groups and Sylow's Theorem are presented in a separate section.

FINITE ABELIAN GROUPS

THEOREM 8.1

If G is a cyclic group of order n and if $m|n$, then G has a subgroup of order m.

Proof: Let $G = \langle g \rangle$ be a cyclic group of order n. If $m|n$, then $n = mk$ for some integer k. Let $h = g^k$. Then h has order m because

$$h^m = (g^k)^m = g^n = e.$$

No smaller power of h can give e since no smaller power of g can give e. Thus $H = \langle h \rangle$ has order m and our result is proved. ■

EXAMPLE 8.1
Consider a cyclic group of order 12, $G = \{e, g, g^2, \ldots, g^{11}\}$. By Theorem 8.1 and its proof, we know that G has subgroups of orders

1: $\{e\}$
2: $\{e, g^6\}$
3: $\{e, g^4, g^8\}$
4: $\{e, g^3, g^6, g^9\}$

and

6: $\{e, g^2, g^4, g^6, g^8, g^{10}\}$.

Theorem 8.1 does not hold for all finite groups. We cannot say, in general, that if $m|n$, then a group of order n has a subgroup of order m. We will give an example to illustrate this fact in Chapter 9. The existence of subgroups of specific orders depends on the order n of the group G and on the primes that divide n. The next theorem gives us one of these criteria.

THEOREM 8.2
If $n = p \cdot q$ where p and q are distinct primes, and if G is an abelian group of order n, then there exists an element of order p and an element of order q in G.

Proof: Choose an element g in G with $g \neq e$, the identity of G. $\langle g \rangle < G$ and therefore, by Lagrange's Theorem, the subgroup $\langle g \rangle$ and, hence, the element g have order either p, q or $p \cdot q$.

Case 1: Suppose g has order $p \cdot q$. Then g^p is the element of order q and g^q is the element of order p as in the proof of the last theorem.

Case 2: Assume that g has order p. We need to show that there exists an element of order q.

Consider the cyclic subgroup $\langle g \rangle < G$. Since G is abelian, $\langle g \rangle$ is a normal subgroup of G and we can look at the quotient group $G|\langle g \rangle$ which is of order $q = (|G|/|\langle g \rangle|)$. Choose an element $\langle g \rangle \circ a$ in $G|\langle g \rangle$. If $\langle g \rangle \circ a \neq \langle g \rangle$ (the identity in $G|\langle g \rangle$), then the order of the element $\langle g \rangle \circ a$ is q since it must divide $|G|\langle g \rangle| = q$ and q is prime. Therefore

$$(\langle g \rangle \circ a)^q = \langle g \rangle \circ a^q = \langle g \rangle$$

or $a^q \in \langle g \rangle$, a group of order p. Therefore

$$(a^q)^p = (a^p)^q = e$$

and a^p will be the desired element of order q if we can prove that $a^p \neq e$. We do this by contradiction.

If $a^p = e$, then

$$(\langle g \rangle \circ a)^p = \langle g \rangle \circ a^p$$
$$= \langle g \rangle \circ e$$
$$= \langle g \rangle, \quad \text{the identity in } G|\langle g \rangle.$$

But $\langle g \rangle \circ a$ was an element of order q and $(\langle g \rangle \circ a)^p$ is the identity, which implies that $p|q$. This contradicts the hypothesis that p and q are distinct primes.

Therefore $a^p \neq e$ so that a^p has order q.

Case 3: Assume that g has order q. This is the same as Case 2 with the roles of p and q interchanged. ∎

Theorem 8.2 is generalized in the next result, Cauchy's Theorem for Abelian Groups. The proof is similar to the proof of Theorem 8.2, but is complicated by the use of induction. It is presented in the last section of this chapter.

THEOREM 8.3: CAUCHY'S THEOREM FOR ABELIAN GROUPS
If G is an abelian group of order n and if p is a prime that divides n, then there is an element of order p in G.

COROLLARY
If G is an abelian group of order n and if p is any prime that divides n, then G has a subgroup of order p.

Proof: If $g \in G$ is the element of G with order p, then its cyclic subgroup $\langle g \rangle$ also has order p. ∎

We take this concept one step further with the next theorem.

THEOREM 8.4: SYLOW'S THEOREM FOR ABELIAN GROUPS
Let G be a finite abelian group of order n and let p be a prime number. If k is a positive integer for which p^k divides n but p^{k+1} does not divide n, then G has a subgroup of order p^k.

Proof: Let G be an abelian group of order n. Let p be a prime such that for some integer $k > 0$, $p^k | n$ and $p^{k+1} \nmid n$. Let

$H = \{h \in G | h^{p^m} = e \text{ for some integer } m \in Z\}$.

We will show (1) that H is a subgroup of G and (2) that $|H| = p^k$.

1. We note that $H \neq \emptyset$; $e^{p^m} = e$ for any integer m so that $e \in H$. Using Theorem 5.2 to prove $H < G$, we choose elements x and $y \in H$ and show

that $x \circ y^{-1} \in H$. Since $x \in H$ and $y \in H$, $x^{p^m} = e$ and $y^{p^l} = e$ for some integers m and l. Therefore

$$\begin{aligned}(x \circ y^{-1})^{p^{m+l}} &= x^{p^{m+l}} \circ (y^{-1})^{p^{m+l}} & \text{(since } G \text{ is abelian)} \\ &= (x^{p^m})^{p^l} \circ ((y^{-1})^{p^l})^{p^m} & \text{(why?)} \\ &= e \circ e & \text{(why?)} \\ &= e.\end{aligned}$$

Therefore $x \circ y^{-1} \in H$ and H is a subgroup of G. Note that since G is abelian, $H \triangleleft G$.

2. To show that $|H| = p^k$ we will first show that for any prime $q \neq p$ that $q \nmid |H|$. This implies that $|H| = p^l$ where $l \leq k$ (why?). We will then show that $l < k$ leads to a contradiction so that $|H| = p^k$.

Suppose q is a prime. If $q \mid |H|$, then, by Cauchy's Theorem for Abelian Groups, there is an element $h \in H$ of order q. Since $h^{p^n} = e$ for some integer n, $q \mid p^n$ (why?). The only prime for which this is true is $q = p$. Therefore $|H| = p^l$ for some integer $l \leq k$.

If $l < k$, p divides the order of the quotient group $G|H$.

$$|G|H| = \frac{|G|}{|H|} = \frac{p^k A}{p^l B}.$$

By Cauchy's Theorem for Abelian Groups, there is an element $H \circ g \in G|H$ of order p.

$$(H \circ g)^p = H \circ g^p = H$$

which implies $g^p \in H$. Therefore, for some integer m, $(g^p)^{p^m} = e$ or $g^{p^{m+1}} = e$ which implies that $g \in H$ or $H \circ g = H$. But this contradicts the fact that $H \circ g$ is of order p. Our false assumption was that $l < k$. We conclude that $|H| = p^k$. ∎

EXAMPLE 8.2
If G is an abelian group of order $2^3 \cdot 5^2 \cdot 13^4$, Cauchy's Theorem guarantees subgroups of orders 2, 5, and 13. By Sylow's Theorem, there are subgroups of orders 2^3, 5^2, and 13^4. Note that from these theorems we have no information about the possible existence of subgroups of orders $2 \cdot 5, 2^2, 5^2 \cdot 13^2, 13^3$.

EXERCISES
1. If G is an abelian group with $|G| = 3^4 \cdot 2^5 \cdot 7^2$, for which orders do Sylow's Theorem and Cauchy's Theorem guarantee subgroups?
2. a. If $G = \{e, g, g^2, \ldots, g^{23}\}$ is a cyclic group, find the subgroup H of the proof of Sylow's Theorem for each of the primes 2 and 3.

b. Find the subgroup H of the proof of Sylow's Theorem for the prime 2 and the Klein four-group.
3. Show that Sylow's Theorem is true for $k = 0$.
4. a. Let G be an abelian group and let

$$G' = \{g \in G | g^n = e \text{ for some integer } n\}.$$

   Prove that $G' < G$.
   b. Is $G' < G$ if G is not abelian?
5. a. Suppose G is an abelian group and n is an integer. Show that $H = \{g \in G | g^n = e\}$ is a subgroup of G.
   b. For each positive integer n find the subgroup H of part a in the Klein four-group.

CENTERS, CENTRALIZERS, AND NORMALIZERS

In a non-abelian group, there must be at least one pair of elements x and y for which $x \circ y \neq y \circ x$. However, we have seen that non-abelian groups have some abelian subgroups and that the identity commutes with every element of the group. We extend this concept by considering the set of elements that commute with all elements of a group.

DEFINITION
If G is a group, let

$$C(G) = \{c \in G | c \circ g = g \circ c \text{ for all } g \in G\}.$$

$C(G)$ is called the **center** of G. An element $c \in C(G)$ is called a **central element** of G.

$C(G)$ is a subset of G. We also see that $C(G)$ is never empty since, for every group G, the identity element is a member of C.

EXAMPLES 8.3
1. We find the center of the group $(N_{2 \times 2}, \cdot)$ where

$$N_{2 \times 2} = \left\{ \begin{pmatrix} a & b \\ c & d \end{pmatrix} \middle| ad - bc \neq 0 \right\}.$$

Let $\begin{pmatrix} a & b \\ c & d \end{pmatrix}$ be any matrix in $N_{2 \times 2}$. $\left(\begin{pmatrix} a & b \\ c & d \end{pmatrix} \text{ is fixed but arbitrary.} \right)$

If $\begin{pmatrix} x & y \\ z & w \end{pmatrix}$ belongs to $C(N_{2 \times 2})$, then

$$\begin{pmatrix} x & y \\ z & w \end{pmatrix} \begin{pmatrix} a & b \\ c & d \end{pmatrix} = \begin{pmatrix} a & b \\ c & d \end{pmatrix} \begin{pmatrix} x & y \\ z & w \end{pmatrix}.$$

But
$$\begin{pmatrix} x & y \\ z & w \end{pmatrix} \begin{pmatrix} a & b \\ c & d \end{pmatrix} = \begin{pmatrix} ax + cy & bx + dy \\ az + cw & bz + dw \end{pmatrix}$$
while
$$\begin{pmatrix} a & b \\ c & d \end{pmatrix} \begin{pmatrix} x & y \\ z & w \end{pmatrix} = \begin{pmatrix} ax + bz & ay + bw \\ cx + dz & cy + dw \end{pmatrix}.$$

Equating the corresponding entries in the two product matrices, we get

$ax + cy = ax + bz$
$bx + dy = ay + bw$
$az + cw = cx + dz$
$bz + dw = cy + dw$

which is a system of linear equations in the unknowns x, y, z, and w. The solution to this system is $a = d$, $b = c = 0$. (You are asked to verify this in the exercises.) Therefore

$$C(N_{2 \times 2}) = \left\{ \begin{pmatrix} a & 0 \\ 0 & a \end{pmatrix} \middle| a \in \mathbb{R} \right\}.$$

2. In the group S_3, the center $C(S_3)$ consists of the identity alone, $C(S_3) = \{f_0\}$. We can see this by looking at the table on p. 26. The first row equals the first column which implies that $f_0 \in C(S_3)$. No other corresponding row and column are equal.

3. In the dihedral group D_4 (Example 3.4) we can see that $C(D_4) = \{I, r_2\}$.

4. If G is an abelian group, for any $x \in G$, $x \circ y = y \circ x$ for all $y \in G$. Therefore every element of G belongs to the center of G or $C(G) = G$.

Notice that in each of these examples the center of the group is a subgroup. We will prove that this is true in general.

THEOREM 8.5

If G is a group, then $C(G) < G$.

Proof: Note that $C(G) \neq \emptyset$. To see that $C(G)$ is a subgroup, let x and y be any two elements of $C(G)$. By Theorem 5.2, it suffices to show that $x \circ y^{-1} \in C(G)$.

Let g be an arbitrary element of G. We will show that $(x \circ y^{-1})$ commutes with g.

First, since $y \circ g = g \circ y$ for any $g \in G$, we can conclude that $g \circ y^{-1} = y^{-1} \circ g$ for any $g \in G$ (why?) or that $y^{-1} \in C(G)$ whenever y does.

The structure of finite groups 101

Using the fact x and y^{-1} are in $C(G)$, we can write

$$(x \circ y^{-1}) \circ g = x \circ g \circ y^{-1} \quad \text{for any } g \in G \quad \text{(why)?}$$
$$= g \circ (x \circ y^{-1})$$

which is the equality we were trying to prove. Since $(x \circ y^{-1}) \circ g = g \circ (x \circ y^{-1})$ for any g in G, we have $x \circ y^{-1} \in C(G)$ for any $x, y \in C(G)$, so $C(G)$ is a subgroup of G. ∎

In the next theorem, we list several elementary properties of the center $C(G)$ of a group G.

THEOREM 8.6
For any group G with center $C(G)$
1. $C(G)$ is an abelian group.
2. $C(G)$ is normal in G.
3. If H is any subgroup of $C(G)$, then $H \triangleleft G$.

Proof of 1: Choose any two elements x and y in $C(G)$. Since x and y commute with all elements of G, they commute with each other or $x \circ y = y \circ x$ and $C(G)$ is abelian.

The proofs of parts 2 and 3 are left as exercises. ∎

When we were discussing the automorphism group of a group G, we defined, in the exercises, the subgroup of inner automorphisms: for $g \in G$, the inner automorphism α_g is given by

$$\alpha_g(x) = g^{-1} \circ x \circ g \quad \text{for all } x \in G.$$

We saw, in those exercises, that $G|C(G)$ is essentially the same as (isomorphic to) the group of inner automorphisms. For some purposes the notation of the center of a group is preferable to that of inner automorphisms.

A related concept is that of the centralizer of a group element.

DEFINITION
If G is a group and $g \in G$, then the **centralizer** of g in G is defined to be

$$C_G(g) = \{x \in G \mid x \circ g = g \circ x\}.$$

That is, the centralizer $C_G(g)$ of g is the set of all elements that commute with g.

EXAMPLE 8.4

In $S_3 = \{f_0, f_1, f_2, f_3, f_4, f_5\}$, the symmetric group on three elements, we will find the centralizer of each element.

$C_{S_3}(f_0) = \{f_0, f_1, f_2, f_3, f_4, f_5\}$
$C_{S_3}(f_1) = \{f_0, f_1\}$
$C_{S_3}(f_2) = \{f_0, f_2\}$
$C_{S_3}(f_3) = \{f_0, f_3, f_4\}$
$C_{S_3}(f_4) = \{f_0, f_3, f_4\}$
$C_{S_3}(f_5) = \{f_0, f_5\}$.

Again we see that each of the above sets is a subgroup of S_3. Again, this is true for all groups.

THEOREM 8.7
If G is a group and $g \in G$, then $C_G(g)$ is a subgroup of G.

The proof is left as an exercise.

The next results deal with subgroups rather than elements, but the concept is similar.

Notation: For a group G and a subgroup $H < G$,

$$g^{-1} \circ H \circ g = \{g^{-1} \circ h \circ g \mid h \in H\}.$$

DEFINITION

Let H be a subgroup of a group G. $N_G(H)$ is defined to be $\{g \in G \mid g^{-1} \circ H \circ g = H\}$. $N_G(H)$ is called the **normalizer** of H.

That is, if g is in the normalizer of H, then every $h \in H$ is of the form $g^{-1} \circ h_1 \circ g$ for some $h_1 \in H$, and for every $h \in H$, $g^{-1} \circ h \circ g \in H$.

EXAMPLES 8.5

1. In S_3, let $H = \{f_0, f_2\}$. To find $N_G(H)$, we need to check the sets $g^{-1} \circ H \circ g$ for every $g \in S_3$:

$f_0^{-1} \circ H \circ f_0 = \{f_0^{-1} \circ f_0 \circ f_0, f_0^{-1} \circ f_2 \circ f_0\} = \{f_0, f_2\} = H$
$f_1^{-1} \circ H \circ f_1 = \{f_0, f_5\} \neq H$
$f_2^{-1} \circ H \circ f_2 = \{f_0, f_2\} = H$
$f_3^{-1} \circ H \circ f_3 = \{f_0, f_1\} \neq H$
$f_4^{-1} \circ H \circ f_4 = \{f_0, f_5\} \neq H$
$f_5^{-1} \circ H \circ f_5 = \{f_0, f_1\} \neq H$.

Therefore $N_G(H) = \{f_0, f_2\}$.

2. For the subgroup $N = \{f_0, f_3, f_4\} < S_3$, we can see that $N_G(H) = S_3$ by performing the same computations. (Note that $N \triangleleft S_3$.)

3. If G is any abelian group and H is any subgroup of G, then $H \triangleleft G$ and $N_G(H) = G$ (Exercise 8).

Our main results about normalizers are given in the next two theorems.

THEOREM 8.8

If G is a group and $H < G$, then $N_G(H)$ is a subgroup of G containing H.

Proof: We will show (1) that $H \subseteq N_G(H)$ and (2) that $N_G(H) < G$.

1. For any $h \in H$, consider the set $h^{-1} \circ H \circ h$. Any element of this set is of the form $h^{-1} \circ h_1 \circ h$ for some $h_1 \in H$, and $h^{-1} \circ h_1 \circ h \in H$; therefore $h^{-1} \circ H \circ h \subseteq H$ for every $h \in H$. Exercise 6 states that it is unnecessary to prove set containment in the other direction. Therefore, for every $h \in H$, $h \in N_G(H)$ or $H \subseteq N_G(H)$.

2. $N_G(H) \neq \emptyset$ since $\emptyset \neq H \subseteq N_G(H)$. Choose elements x and y in $N_G(H)$; then $x^{-1} \circ H \circ x = H$ and $y^{-1} \circ H \circ y = H$. This second equality implies that $y \circ H \circ y^{-1} = H$ also.

We want to show that $x \circ y^{-1} \in N_G(H)$ or that $(x \circ y^{-1})^{-1} \circ H \circ (x \circ y^{-1}) = H$:

$(x \circ y^{-1})^{-1} \circ H \circ (x \circ y^{-1}) = (y \circ x^{-1}) \circ H \circ (x \circ y^{-1})$ (why?)

$\qquad = y \circ H \circ y^{-1}$ (why?)

$\qquad = H$.

Therefore $x \circ y^{-1} \in H$ for any x and $y \in N_G(H)$ and by Theorem 5.2, $N_G(H) < G$. ∎

THEOREM 8.9

If G is a group, $H < G$ and $N_G(H)$, the normalizer of H in G, then $H \triangleleft N_G(H)$ and $N_G(H)$ is the largest subgroup for which this is true.

Proof: We need to show that (1) H is normal in $N_G(H)$ and (2) for any subgroup K such that $H \triangleleft K < G$, then $K \subseteq N_G(H)$.

1. For all $g \in N_G(H)$, $g^{-1} \circ H \circ g = H$ and we have shown in an exercise in Chapter 6 that this is equivalent to the statement H is normal in $N_G(H)$.

2. If $K < G$ and $H \triangleleft K$, then for every $k \in K$, $k^{-1} \circ H \circ k = H$ (why?) and $k \in N_G(H)$. Therefore $K \subseteq N_G(H)$. ∎

EXERCISES

1. a. Prove that for any group G, $C(G)$ is normal in G.
   b. Suppose $H < C(G)$; prove that H is normal in G.
2. If G is a group and g is any element of G, prove that $C_G(g) < G$ and $g \in G$.

3. Show that $x \in C_G(g)$ if and only if $g = x^{-1} \circ g \circ x$.
4. Prove that $x \in C_G(g)$ if and only if $g \in C_G(x)$.
5. a. Solve the system of equations in the first example in this subsection.
   b. Show directly that $C(N_{2 \times 2})$ is a subgroup of $N_{2 \times 2}$.
6. Let G be a finite group and $H < G$. Prove that $g^{-1} \circ H \circ g = H$ if and only if $g^{-1} \circ H \circ g \subseteq H$.
7. If G is abelian and $H < G$, show that $N_G(H) = G$.
8. Show that $y^{-1} \circ H \circ y = H$ implies $H = y \circ H \circ y^{-1}$.
9. For the proofs in this chapter supply the reasons that were omitted.
10. a. In the dihedral group D_4, find the normalizers of the subgroups $\{I, r_2\}, \{I, r_1, r_2, r_3\}, \{I, d_1\}$.
    b. Find the centralizers of the elements r_1, r_2, h, d_2.

CONJUGACY

The last concept we have to develop before we present the proofs to Cauchy's Theorems and Sylow's Theorem is that of conjugate classes in a group. We have already explored the inner automorphism α_g defined by $\alpha_g(x) = g^{-1} \circ x \circ g$. Here we look at the same thing in a slightly different way.

DEFINITION

If G is a group and if x and $y \in G$, then if for some $g \in G$, $y = g^{-1} \circ x \circ g$, we say that y is **conjugate** to x.

For every $x \in G$, we define $C_x = \{y \in G \mid y \text{ is conjugate to } x\}$. C_x is called the **conjugate class** of x.

Notation: If G is a finite group, we can write $G = \{e, x_1, \ldots, x_n\}$. We will denote the conjugate class of the element $x_i \in G$, $i = 1, \ldots, n$, by C_i.

We wish to prove that the sets of elements which are conjugate to each other fall into mutually disjoint classes of elements of the group G. We would like to write G as a set union, $G = C_1 \cup C_2 \cup \cdots \cup C_m$, where $C_i \cap C_j = \emptyset$ if $i \neq j$. That is, we want to show that the classes of all conjugate elements form a partition of G. One way of doing this is to prove that conjugacy is an equivalence relation.

THEOREM 8.10

If G is a group, and the relation r on G is defined by $x \text{ r } y$ means x is conjugate to y, then r is an equivalence relation.

Proof: We have to prove that the relation is (1) reflexive, (2) symmetric, and (3) transitive.

1. For any $x \in G$, x is conjugate to itself since there is an identity $e \in G$ with $x = e^{-1} \circ x \circ e$. The relation is reflexive.

2. If x is conjugate to y, then there exists an element $g \in G$ such that $x = g^{-1} \circ y \circ g$. Therefore $y = (g^{-1})^{-1} \circ x \circ g^{-1}$ or y is conjugate to x and the relation is symmetric.

3. If x is conjugate to y and y is conjugate to z, then there are elements g and h in G with the property that $x = g^{-1} \circ y \circ g$ and $y = h^{-1} \circ z \circ h$. Therefore, by substitution,

$$x = g^{-1} \circ (h^{-1} \circ z \circ h) \circ g$$
$$= (g^{-1} \circ h^{-1}) \circ z \circ (h \circ g)$$
$$= (h \circ g)^{-1} \circ z \circ (h \circ g)$$

or x is conjugate to z and the relation is transitive.

Since the relation of conjugacy satisfies all three properties—reflexive, symmetric, and transitive—it is an equivalence relation. ∎

COROLLARY

The set of all conjugate classes, $\{C_x \mid x \in G\}$, forms a partition of G.

Proof: This follows immediately from Theorem 6.1 and the fact that conjugacy is an equivalence relation. ∎

EXAMPLE 8.6

In S_3 let us find the conjugacy classes. To find $C_0 = C_{f_0}$ we need to consider the 6 products:

$f_0^{-1} \circ f_0 \circ f_0 = f_0 \qquad f_3^{-1} \circ f_0 \circ f_3 = f_0$
$f_1^{-1} \circ f_0 \circ f_1 = f_0 \qquad f_4^{-1} \circ f_0 \circ f_4 = f_0$
$f_2^{-1} \circ f_0 \circ f_2 = f_0 \qquad f_5^{-1} \circ f_0 \circ f_5 = f_0.$

Therefore $C_0 = \{f_0\}$. In fact, we can see that the conjugate class of the identity in any group G will always be just the identity.

To find C_1 we compute

$f_0^{-1} \circ f_1 \circ f_0 = f_1 \qquad f_3^{-1} \circ f_1 \circ f_3 = f_5$
$f_1^{-1} \circ f_1 \circ f_1 = f_1 \qquad f_4^{-1} \circ f_1 \circ f_4 = f_2$
$f_2^{-1} \circ f_1 \circ f_2 = f_5 \qquad f_5^{-1} \circ f_1 \circ f_5 = f_2$
$C_1 = \{f_1, f_2, f_5\}.$

Similarly, we find

$C_2 = \{f_2, f_1, f_5\} = C_1 = C_5$
$C_3 = \{f_3, f_4\} = C_4.$

Our first result on conjugate classes is the following lemma.

LEMMA

If $G = \{x_1, x_2, \ldots, x_n\}$, then C_i consists of the single group element x_i if and only if x_i is in the center of G.

Proof: Assume $x_i \in C(G)$. Then, for every $g \in G$, $g^{-1} \circ x_i \circ g = x_i$ (why?) and x_i is conjugate only to itself; therefore $C_i = \{x_i\}$.

Assume $C_i = \{x_i\}$. Then, for all $g \in G$, $g^{-1} \circ x_i \circ g = x_i$ or $x_i \circ g = g \circ x_i$ and $x_i \in C(G)$. ∎

Now let G be a group of order n and let $C(G)$ be a subgroup of order k (k divides n). There are k singleton conjugate classes which we denote by $C_1, C_2, \ldots, C_k$. We let $C_{k+1}, C_{k+2}, \ldots, C_m$ be the remaining classes. We can write

$$G = C_1 \cup C_2 \cup \cdots \cup C_k \cup C_{k+1} \cup \cdots \cup C_m.$$

If we let $h_i = $ the number of elements in C_i for $i = 1, 2, \ldots, m$, then we get

$$(*) \quad n = \underbrace{1 + 1 + \cdots + 1}_{k} + h_{k+1} + h_{k+2} + \cdots + h_m$$

$$= k + h_{k+1} + \cdots + h_m.$$

Equation (*) is called the **class equation** of G.

EXAMPLE 8.7

The class equation for S_3 is

$$6 = 1 + 3 + 2.$$

We need one final result before proving Cauchy's Theorems and Sylow's Theorem. Recall that $[G:H]$ is called the index of H in G and equals the number of cosets of H in G.

THEOREM 8.11

Let G be a finite group, $|G| = n$, with conjugate classes $C_1, C_2, \ldots, C_m$ as above, and class equation

$$n = k + h_{k+1} + h_{k+2} + \cdots + h_m.$$

Then for any group element $x_i \in C_i$, it is true that

$$h_i = [G : C_G(x_i)]$$

the index of the centralizer of x_i in G.

Proof: Case 1: If $x_i \in C(G)$, then $C_i = \{x_i\}$ and $h_i = 1$. Also, $C_G(x_i) = G$ and $[G : C_G(x_i)] = 1$, verifying the theorem for this case.

The structure of finite groups

Case 2: If $x_i \notin C(G)$, then $C_G(x_i) \neq G$. To simplify our notation, let $H_i = C_G(x_i)$. We shall set up a bijection (1–1 correspondence) between the right cosets of H_i in G and the elements of C_i, verifying our assertion that the number of cosets (index) is equal to the number of elements in C_i. Let $\mathcal{H} = \{H_i \circ a \mid a \in G\}$, the set of right cosets of H_i in G. Recall $C_i = \{a^{-1} \circ x_i \circ a \mid a \in G\}$. We define

$$\psi: \mathcal{H} \to C_i$$

by

$$\psi(H_i \circ a) = a^{-1} \circ x_i \circ a.$$

We need to show that ψ is (1) well defined, (2) injective, and (3) surjective.

1. Choose $c \in H_i \circ a$; $H_i \circ a = H_i \circ c$ and we need to show that $\psi(H_i \circ a) = \psi(H_i \circ c)$. Since $c \in H_i \circ a$, $c = h \circ a$ for some $h \in H_i = C_G(x_i)$,

$$\begin{aligned}
\psi(H_i \circ c) &= c^{-1} \circ x_i \circ c \\
&= (h \circ a)^{-1} \circ x_i \circ (h \circ a) \\
&= a^{-1} \circ h^{-1} \circ x_i \circ h \circ a \quad \text{(why?)} \\
&= a^{-1} \circ h^{-1} \circ h \circ x_i \circ a \quad \text{(since } h \in C_G(x_i)\text{)} \\
&= a^{-1} \circ x_i \circ a \\
&= \psi(H_i \circ a).
\end{aligned}$$

Since different coset representatives yield the same image elements, ψ is well defined.

2. Suppose $a^{-1} \circ x_i \circ a = b^{-1} \circ x_i \circ b$. To show that ψ is injective, we must show that $H_i \circ a = H_i \circ b$.

$$\begin{aligned}
a^{-1} \circ x_i \circ a &= b^{-1} \circ x_i \circ b \\
\Rightarrow \quad (b \circ a^{-1}) \circ x_i &= x_i \circ (b \circ a^{-1}) \quad \text{(why?)} \\
\Rightarrow \quad b \circ a^{-1} &\in C_G(x_i) = H_i \quad \text{(why?)}.
\end{aligned}$$

Therefore $b \in H_i \circ a$ or $H_i \circ b = H_i \circ a$.

3. Choose any element $a^{-1} \circ x_i \circ a \in C_i$. Since $\psi(H_i \circ a) = a^{-1} \circ x_i \circ a$, ψ is surjective and we have completed the proof. ∎

This theorem gives us the last material necessary in the proofs of Cauchy's Theorems and Sylow's Theorem. Exhibiting a bijection between two sets (as was done in the proof) to show that the two sets have the same number of elements is a common algebraic technique. To repeat, the theorem says that if the class equation of G is

$$n = k + h_{k+1} + \cdots + h_m$$

then $h_i = [G : C_G(x_i)]$ where $x_i \in C_i$.

EXERCISES

1. Find the conjugate classes and class equation for the dihedral group D_4.
2. If e is the identity of any group G, show that the conjugate class of e is $\{e\}$.
3. If G is a cyclic group of order 6, what are the conjugate classes?
4. a. If $G = \{x_1, x_2, \ldots, x_n\}$ is abelian, show that $C_i = \{x_i\}$ for $i = 1, 2, \ldots, n$.
   b. What is the class equation of any abelian group of order n?
5. Prove directly, without using Theorem 8.10, that the classes of all conjugate elements partition any group.

PROOFS OF CAUCHY'S THEOREMS
AND SYLOW'S THEOREM (OPTIONAL)

We conclude this chapter with the proofs of Cauchy's Theorem for Abelian Groups, Cauchy's Theorem, and Sylow's Theorem for non-abelian groups.

The proofs of all of these theorems are done by induction. The inductive pattern is to prove the theorem true for any group G with $|G| = 1$; assume it is true for all groups G with $|G| < n$; and prove that the theorem is true for any group G with $|G| = n$. Then the theorem will be true for finite groups of all orders.

CAUCHY'S THEOREM FOR ABELIAN GROUPS

If G is an abelian group of order n and if p is a prime that divides n, then there is an element of order p in G.

Proof: We will proceed by induction.

Suppose $|G| = 1$. Then there is no prime which divides $|G|$. The theorem is true since it is impossible to get a counterexample.

Although it is not necessary, we present the case $|G| = 2$. In this case G is cyclic and the element g has order 2.

Assume the theorem is true for all groups of orders less than n. We will show that the theorem is true for any group G with $|G| = n$.

Suppose G is a finite abelian group with $|G| = n$ and suppose p is a prime that divides n. Choose $g \in G$, $g \neq e$.

Case 1: If the order of g is pm (a multiple of p), then the element g^m has order p as in the proof of Theorem 8.1.

The structure of finite groups 109

Case 2: Suppose the order of g is r and $p \nmid r$. The cyclic group $\langle g \rangle$ also has order r. Since G is abelian, $\langle g \rangle \triangleleft G$ and the quotient group $G/\langle g \rangle$ has order n/r which is less than n. Since $p \nmid r$ and $p \mid n$, $p \mid n/r$ and by the induction hypothesis, there is an element $\langle g \rangle \circ a \in G/\langle g \rangle$ of order p:

$$(\langle g \rangle \circ a)^p = \langle g \rangle \circ a^p = \langle g \rangle.$$

Hence $a^p \in \langle g \rangle$ and therefore $a^{pr} = e$ (why?) and we see that the element a^r has order p unless $a^r = e$.

But if $a^r = e$, then $(\langle g \rangle \circ a)^r = \langle g \rangle \circ a^r = \langle g \rangle$, the identity of $G/\langle g \rangle$. But $\langle g \rangle \circ a$ was an element of order p and therefore $p \mid r$, which is a contradiction.

Therefore a^r has order p and the theorem is proved. ■

The proofs of the next two theorems use the theory of centers, centralizers, and conjugacy as well as induction. Their proofs are very similar.

CAUCHY'S THEOREM

If G is a group of order n and if $p \mid n$, then there is an element of G of order p.

Proof: Again we proceed by induction. If $|G| = 1$ or 2, then the theorem is true for the same reasons as in the last proof.

Assume that Cauchy's Theorem holds for all groups with $|G| < n$.

Let G be a group of order n. If G is abelian, the last theorem holds and Cauchy's Theorem is true.

Assume that G is not abelian and assume that the center of G has order $|C(G)| = k < n$ so that G has class equation

$$n = k + h_{k+1} + h_{k+2} + \cdots + h_m$$

with $h_i = [G : C_G(x_i)]$.

Case 1: Assume there exists an $x \in G$, $x \notin C(G)$, such that $p \mid |C_G(x)|$. Since $|C_G(x)| < n$, the induction hypothesis implies that there is an element $y \in C_G(x)$ of order p. The theorem is true since $C_G(x)$ is a subgroup of G.

Case 2: Assume for all $x_i \in G$, $x_i \notin C(G)$ that $p \nmid |C_G(x_i)|$. Since $|G| = [G : C_G(x_i)] \cdot |C_G(x_i)| = h_i |C_G(x_i)|$ by Theorem 8.11, $p \mid h_i$ for $i = k+1, \ldots, m$. In the class equation of G,

$$n = k + h_{k+1} + \cdots + h_m$$

p divides n, $h_{k+1}, h_{k+2}, \ldots, h_m$ and therefore $p \mid k$. The center $C(G)$ is an abelian group with $p \mid |C(G)|$ and by Cauchy's Theorem for Abelian Groups, there is an element of order p. ■

SYLOW'S THEOREM

Let G be a finite group of order n and let p be a prime number. If for $r > 0$, p^r divides n and p^{r+1} does not divide n, then G has a subgroup of order p^r.

Proof: The proof is done by induction. If $n = 1$ (or 2), G is cyclic and by Theorem 8.1, Sylow's Theorem is true.

Assume that the theorem is true for all groups with order $|G| < n$. We will show that it is true for a group G with order n.

Let G be a group of order n.

If G is abelian, Theorem 8.4 holds and the theorem is true.

If G is not abelian, the order of the center $C(G)$ is less than n. Suppose $|C(G)| = k < n$. The class equation of G is

$$n = k + h_{k+1} + h_{k+2} + \cdots + h_m$$

where $h_j > 1$ for $k + 1 \leq j \leq m$. We will consider the cases (1) $p|k$ and (2) $p \nmid k$ separately.

1. If p divides k, then p^t divides k and p^{t+1} does not divide k for some t, $1 \leq t \leq r$.

    a. If $t = r$, that is, if p^r divides k, then, since $k < n$, $C(G)$ has a subgroup P of order p^r (why?) and we are done.

    b. If $t < r$, then $C(G)$ has a subgroup P of order p^t. Since $P < C(G)$, $P \triangleleft G$ (why?) and we may consider the group $G|P$.

$$|G|P| = [G : P] = \frac{n}{p^t} \quad \text{(why?)}.$$

Thus p^{r-t} divides the order of $G|P$ but p^{r-t+1} does not. Since $n/p^t < n$, we know that $G|P$ satisfies Sylow's Theorem and has a subgroup $H|P$ of order p^{r-t}. Now Theorem 7.12 tells us that G has a subgroup H containing P and corresponding to $H|P$ in $G|P$. Therefore

$$|H| = [H : P] \cdot |P|$$
$$= p^{r-t} \cdot p^t = p^r.$$

Therefore G has to satisfy Sylow's Theorem.

2. If p does not divide k, we rewrite the class equation of G as

$$n - h_{k+1} - h_{k+2} - \cdots - h_m = k.$$

If p divides n and each of $h_{k+1}, \ldots, h_m$, then p would have to divide k also. Therefore, for some h_t, $k + 1 \leq t \leq m$, $p \nmid h_t$. $h_t = [G : C_G(g_t)]$, and $|G| = h_t \cdot |C_G(g_t)|$. Therefore $p^r | |C_G(g_t)|$. By our assumption, since $|C_G(g_t)| < n$, it must have a subgroup P of order p^r and $P < G$ also. ∎

9 Permutation groups

In this chapter we introduce two important topics in group theory: permutation groups and group solvability. Both of these topics are essential in a first course in group theory. We will also find them very important when we are studying fields.

In Chapter 3 we investigated the group $(S_3, \circ)$, the symmetric group on three elements. We considered the set S_3 of all bijections $f_i: X_3 \to X_3$, $i = 0, 1, \ldots, 5$. We defined the operation $\circ$ of composition of functions of S_3 and showed that the system $(S_3, \circ)$ is a group. In this chapter we will do the same thing in general and study the groups in more depth.

DEFINITION
For any positive integer n, let $X_n = \{1, 2, \ldots, n\}$. A **permutation** s on X_n is a bijection $s: X_n \to X_n$. Let S_n be the set of all permutations of X_n.

EXAMPLES 9.1
1. For $n = 1$, $X_1 = \{1\}$. The only permutation on X_1 (bijection from X_1 to itself) is the function s defined by

$s(1) = 1$.

Therefore $S_1 = \{s\}$.

2. For $n = 2$, $X_2 = \{1, 2\}$. We can define permutations s_1 and s_2 on X_2 by

$s_1(1) = 1$ and $s_2(1) = 2$
$s_1(2) = 2$ $s_2(2) = 1$

$S_2 = \{s_1, s_2\}$.

3. For $n = 3$, $X_3 = \{1, 2, 3\}$. S_3 is the set of six functions described in Chapter 3.

Our next goal will be to define a group operation on S_n. As in Chapter 3, we will introduce the operation of composition of functions. If $f: X_n \to X_n$ and $g: X_n \to X_n$ are permutations of X_n, then we define the composition $f \circ g$ as follows: For any element $i \in X_n$,

$$(f \circ g)(i) = f(g(i)).$$

LEMMA

$\circ$ is a binary operation on S_n.

Proof: For $f, g \in S_n$, we need to show that $f \circ g \in S_n$; that is, we need to show that the function $f \circ g: X_n \to X_n$ is (1) injective and (2) surjective.

1. To show $f \circ g$ is injective, choose $i, j \in X_n$ with $i \neq j$. Since $g: X_n \to X_n$ is a bijection and therefore injective, $g(i), g(j) \in X_n$ and $g(i) \neq g(j)$. Since f is also injective, $f(g(i))$ and $f(g(j))$ are in X_n and $f(g(i)) \neq f(g(j))$.

Therefore $i \neq j \Rightarrow (f \circ g)(i) \neq (f \circ g)(j)$ and $f \circ g$ is injective.

2. To show $f \circ g$ is surjective, choose an element $k \in X_n$. Since f is surjective (f is bijective), there is an element $j \in X_n$ with $f(j) = k$. Also g is surjective which implies that there is an element $i \in X_n$ with $g(i) = j$. We see that for any element $k \in X_n$, there is an element $i \in X_n$ with

$$k = (f \circ g)(i) = f(g(i))$$

and $f \circ g$ is surjective.

Thus $f \circ g \in S_n$ since it is bijective. ∎

THEOREM 9.1

$(S_n, \circ)$ is a group.

The proof of this theorem is almost identical with that of Theorem 7.4 and is left as an exercise. Note that the elements of S_n are not isomorphisms and that part of the proof of Theorem 7.4 will not apply in this case.

DEFINITION

The group $(S_n, \circ)$ is called the **symmetric group of degree n**. Any subgroup of S_n is called a **permutation group of degree n**.

How many elements are in S_n? That is, how many permutations of X_n are there? We saw that S_1 has one element; S_2 has two; and S_3 has six. In general, what is the order of S_n? The following theorem answers this question.

THEOREM 9.2

If $X_n = \{1, 2, \ldots, n\}$ and S_n is the group of all permutations of X_n, then $|S_n| = n! = n(n-1) \cdot \cdots \cdot 2 \cdot 1$.

Note: The notation $n!$ denotes the product of the first n positive integers. Note that $1! = 1$, $2! = 2$, $3! = 6$, and $4! = 24$.

Proof: We count all the possible permutations in S_n. A permutation of X_n can take the integer 1 to any of the n integers in X_n. Each choice would give a different permutation. The integer 2 could be mapped into any of the $n - 1$ remaining elements in X_n. Therefore we have $n(n - 1)$ possibilities for the images of the integers 1 and 2 and each possibility would give a different permutation.

There are $(n - 2)$ choices for the image of 3 and $(n - 3)$ choices for the image of 4. We continue in the same manner.

Finally, after the choices of the images of the elements $1, 2, \ldots, n - 1$, there is exactly one remaining element for the image of n.

Totally, there are $n(n - 1)(n - 2) \cdot \cdots \cdot (1) = n!$ different permutations in S_n. ∎

One of the properties of permutation groups is that they include all finite groups in some reasonable sense. Cayley's Theorem (Theorem 9.3) demonstrates this fact. We will prove two lemmas that are used in the proof of Cayley's Theorem.

Let $G = \{g_1, g_2, \ldots, g_n\}$ be any group of order n. For any element $x \in G$, define a mapping $l_x : G \to G$ by

$$l_x(g_i) = x \circ g_i.$$

LEMMA 1

$l_x : G \to G$ is a bijection.

Proof: We need to show that l_x is (1) injective and (2) surjective.

1. To show that l_x is injective, choose $g_i, g_j \in G$ with $g_i \neq g_j$. We will show that $l_x(g_i) \neq l_x(g_j)$ by contradiction. Assume $l_x(g_i) = l_x(g_j)$. By definition

$$x \circ g_i = x \circ g_j$$

and by the cancellation laws,

$$g_i = g_j$$

which contradicts our initial choice. Therefore $g_i \neq g_j$ implies $l_x(g_i) \neq l_x(g_j)$ and l_x is injective.

2. To show l_x is surjective, choose any $g_i \in G$ and find an element of G which maps to g_i under l_x. The element $x^{-1} \circ g_i$ has this property.

Since l_x is injective and surjective, it is bijective. ∎

An intuitive approach to the meaning of this lemma is that in a group table for any finite group G, the row of the table corresponding to x would represent the image of G under l_x. We have seen previously in an exercise set in Chapter 6 that this row is a rearrangement of elements of G.

Consider the composition of maps l_x and l_y. For any $g \in G$,

$$(l_x \circ l_y)(g) = l_x(l_y(g))$$
$$= l_x(y \circ g)$$
$$= x \circ (y \circ g)$$
$$= (x \circ y) \circ g$$
$$= l_{x \circ y}(g)$$

or $l_x \circ l_y = l_{x \circ y}$.

For $x \in G$, l_x is a bijection of G onto itself. If we define a function $p: X_n \to G$ by $p(i) = g_i$, then p is obviously a bijection (Exercise 3). By the proof of Theorem 7.4, p^{-1} is also a bijection and the composition of bijections is a bijection. Consider the function

$$\alpha_x = p^{-1} \circ l_x \circ p : X_n \to X_n$$

α_x is a bijection from X_n to X_n or $\alpha_x \in S_n$.

The composition of α_x and α_y is given by

$$\alpha_x \circ \alpha_y = p^{-1} \circ l_x \circ p \circ p^{-1} \circ l_y \circ p$$
$$= p^{-1} \circ l_x \circ l_y \circ p \qquad \text{(why?)}$$
$$= p^{-1} \circ l_{x \circ y} \circ p \qquad \text{(why?)}$$
$$= \alpha_{x \circ y}.$$

LEMMA 2

Let $L = \{\alpha_x | x \in G\}$; then L is a subgroup of S_n.

Proof: Note that $L \subseteq S_n$ since it is a set of permutations of X_n. Also, since G is a group, $e \in G$ and $\alpha_e \in L$ so that $L = \emptyset$.

Again we will use Theorem 5.2 to prove $L < S_n$. Choose elements α_x and α_y in L. We will show that (1) $(\alpha_y)^{-1} = \alpha_{y^{-1}}$ and (2) $\alpha_x \circ (\alpha_y)^{-1} \in L$.

1. Note that for any integer $i \in X_n$, $\alpha_e(i) = i$ (why?) so that α_e is the identity map. Since

$$\alpha_y \circ \alpha_{y^{-1}} = \alpha_{y \circ y^{-1}} = \alpha_e$$

$\alpha_{y^{-1}}$ is the inverse of α_y.

2. We compute $\alpha_x \circ \alpha_{y^{-1}} = \alpha_{x \circ y^{-1}} \in L$ since the element $x \circ y^{-1} \in G$ whenever x and y are in G. ∎

Permutation groups

THEOREM 9.3: CAYLEY'S THEOREM

If G is a group and $|G| = n$, then G is isomorphic with a subgroup of S_n.

Proof: Consider the map $\psi: G \to L = \{\alpha_x \mid x \in G\}$ defined by

$$\psi(g) = \alpha_g.$$

To show that ψ is an isomorphism, we will show that ψ is (1) injective, (2) surjective, and (3) preserves the operations on the groups.

1. Choose two elements $g, h \in G$ with $g \neq h$. We will show that $\alpha_g \neq \alpha_h$ by contradiction. Assume

$$\alpha_g = \alpha_h$$
$$p^{-1} \circ l_g \circ p = p^{-1} \circ l_h \circ p$$

or

$$l_g = l_h \quad \text{(why?)}.$$

Therefore, for any $x \in G$,

$$l_g(x) = l_h(x)$$

or

$$g \circ x = h \circ x.$$

By the cancellation laws $g = h$ which contradicts our original choice. Therefore $g \neq h$ implies $\alpha_g \neq \alpha_h$ and the function ψ is injective.

2. ψ is surjective since for every $\alpha_g \in L$, ψ maps the group element g to α_g.

3. For elements x and y in G,

$$\psi(x) \circ \psi(y) = \alpha_x \circ \alpha_y = \alpha_{x \circ y} = \psi(x \circ y)$$

and the operation is preserved.

Since ψ is bijective and preserves the group operation, the groups G and L are isomorphic. ∎

The importance of Cayley's Theorem is that by studying the symmetric groups we are furthering our knowledge of all finite groups.

EXERCISES

1. a. If $X_n = \{1, 2, \ldots, n\}$, show that a mapping $\alpha: X_n \to X_n$ is injective if and only if it is surjective.
   b. Find examples to show that a mapping from an infinite set to itself can be either injective or surjective without being both.

2. Prove Theorem 9.1 by proving the following facts.
   a. Show that S_n is associative under the operation of composition.
   b. Find the identity of S_n.
   c. For a function $s \in S_n$,
      (i) define s^{-1}.
      (ii) show that $s^{-1} \in S_n$ (that is, show that s^{-1} is a permutation of X_n).
3. If $G = \{g_1, \ldots, g_n\}$ is a finite group, show that the function $p: X_n \to G$ defined by $p(i) = g_i$ is bijection.
4. a. Use the proofs of the lemmas and Cayley's Theorem to find a subgroup of S_4 that is isomorphic to the Klein four-group.
   b. Do the same for the cyclic group of order 4.
5. Let S be any set. Let $A(S)$ be the set of all bijections from S to itself. Show that any group G is isomorphic to a subgroup of $A(G)$. That is, prove that Cayley's Theorem is true for infinite groups. The proof parallels the one given in the text for finite groups.

Before studying S_n more intensively, we will develop some useful notation for representing and working with elements of S_n. As a model, recall our treatment of the group S_3 in Chapter 3.

A typical element of S_n is defined by its action on each of the integers $1, 2, \ldots, n$. We will represent the permutation s by the notation

$$s = \begin{pmatrix} 1 & 2 & \ldots & n \\ s(1) & s(2) & \ldots & s(n) \end{pmatrix}.$$

EXAMPLES 9.2

1. In S_1 the function s would be represented as $s = \begin{pmatrix} 1 \\ 1 \end{pmatrix}$.

2. In S_2 we represent

$$s_1 = \begin{pmatrix} 1 & 2 \\ 1 & 2 \end{pmatrix} \quad \text{and} \quad s_2 = \begin{pmatrix} 1 & 2 \\ 2 & 1 \end{pmatrix}.$$

3. In S_3 the six functions are represented on page 25.

4. $X_4 = \{1, 2, 3, 4\}$ and $S_4 = $ the set of all permutations of X_4. The functions s and t defined by

$s(1) = 3$ and $t(1) = 4$
$s(2) = 1$ $t(2) = 3$
$s(3) = 2$ $t(3) = 1$
$s(4) = 4$ $t(4) = 2$

are elements of S_4. In the new notation we write

$$s = \begin{pmatrix} 1 & 2 & 3 & 4 \\ 3 & 1 & 2 & 4 \end{pmatrix} \quad \text{and} \quad t = \begin{pmatrix} 1 & 2 & 3 & 4 \\ 4 & 3 & 1 & 2 \end{pmatrix}.$$

Permutation groups

Note that these functions are treated entirely differently than matrices and should not be confused with them.

We can use the notation we just developed to compute the composition of functions in S_n. Suppose that f and g are elements of S_n:

$$f = \begin{pmatrix} 1 & 2 & \cdots & n \\ f(1) & f(2) & \cdots & f(n) \end{pmatrix}, \quad g = \begin{pmatrix} 1 & 2 & \cdots & n \\ g(1) & g(2) & \cdots & g(n) \end{pmatrix}.$$

To compute $f \circ g$, write the representations of f and g next to each other:

$$f \circ g = \begin{pmatrix} 1 & 2 & \cdots & n \\ f(1) & f(2) & \cdots & f(n) \end{pmatrix} \circ \begin{pmatrix} 1 & 2 & \cdots & n \\ g(1) & g(2) & \cdots & g(n) \end{pmatrix}.$$

Assume i is an integer, $1 \leq i \leq n$, and assume $g(i) = j$. To compute the image of the integer i under the composition $f \circ g$, follow $i \to g(i)$ in the representation for g and then $j = g(i) \to f(j) = (f \circ g)(i)$ in the representation for f. That is,

$$(f \circ g) = \begin{pmatrix} 1 & 2 & \cdots & j & \cdots & n \\ f(1) & f(2) & \cdots & f(j) & \cdots & f(n) \end{pmatrix} \circ \begin{pmatrix} 1 & 2 & \cdots & i & \cdots & n \\ g(1) & g(2) & \cdots & g(i) & \cdots & g(n) \end{pmatrix}.$$

For each integer k, $1 \leq k \leq n$, we can compute $(f \circ g)(k)$ the same way. We illustrate with an example.

EXAMPLE 9.3

Consider the composition of the elements

$$\begin{pmatrix} 1 & 2 & 3 & 4 \\ 3 & 1 & 2 & 4 \end{pmatrix} \quad \text{and} \quad \begin{pmatrix} 1 & 2 & 3 & 4 \\ 4 & 3 & 1 & 2 \end{pmatrix}$$

of S_4. The arrows indicate the computation of the image of the integer 2.

$$\begin{pmatrix} 1 & 2 & 3 & 4 \\ 3 & 1 & 2 & 4 \end{pmatrix} \circ \begin{pmatrix} 1 & 2 & 3 & 4 \\ 4 & 3 & 1 & 2 \end{pmatrix} = \begin{pmatrix} 1 & 2 & 3 & 4 \\ 4 & 2 & 3 & 1 \end{pmatrix}.$$

Also

$$\begin{pmatrix} 1 & 2 & 3 & 4 \\ 4 & 3 & 1 & 2 \end{pmatrix} \circ \begin{pmatrix} 1 & 2 & 3 & 4 \\ 3 & 1 & 2 & 4 \end{pmatrix} = \begin{pmatrix} 1 & 2 & 3 & 4 \\ 1 & 4 & 3 & 2 \end{pmatrix}.$$

Since these two results are not equal, we see that $(S_4, \circ)$ is not an abelian group.

EXERCISES

1. Perform the following compositions:

   a. $\begin{pmatrix} 1 & 2 & 3 & 4 \\ 1 & 4 & 2 & 3 \end{pmatrix} \circ \begin{pmatrix} 1 & 2 & 3 & 4 \\ 3 & 1 & 4 & 2 \end{pmatrix}$

   b. $\begin{pmatrix} 1 & 2 & 3 & 4 & 5 \\ 5 & 1 & 4 & 3 & 2 \end{pmatrix} \circ \begin{pmatrix} 1 & 2 & 3 & 4 & 5 \\ 2 & 1 & 5 & 3 & 4 \end{pmatrix}$

   c. $\begin{pmatrix} 1 & 2 & 3 & 4 \\ 4 & 1 & 2 & 3 \end{pmatrix}^2$

   d. $\begin{pmatrix} 1 & 2 & 3 & 4 & 5 & 6 \\ 3 & 2 & 6 & 5 & 4 & 1 \end{pmatrix} \circ \begin{pmatrix} 1 & 2 & 3 & 4 & 5 & 6 \\ 6 & 1 & 3 & 2 & 5 & 4 \end{pmatrix}^2$.

2. a. Find the inverses of the following permutations:

   (i) $\begin{pmatrix} 1 & 2 & 3 \\ 3 & 1 & 2 \end{pmatrix}$

   (ii) $\begin{pmatrix} 1 & 2 & 3 & 4 \\ 4 & 1 & 2 & 3 \end{pmatrix}$

   (iii) $\begin{pmatrix} 1 & 2 & 3 & 4 & 5 & 6 \\ 2 & 1 & 3 & 5 & 6 & 4 \end{pmatrix}$.

   b. Generalize the results from part a to prove that
   $$\begin{pmatrix} 1 & 2 & \cdots & n \\ f(1) & f(2) & \cdots & f(n) \end{pmatrix}^{-1} = \begin{pmatrix} f(1) & f(2) & \cdots & f(n) \\ 1 & 2 & \cdots & n \end{pmatrix}.$$

3. a. Find the orders of the permutations
   $\begin{pmatrix} 1 & 2 & 3 \\ 1 & 3 & 2 \end{pmatrix}$ and $\begin{pmatrix} 1 & 2 & 3 \\ 2 & 3 & 1 \end{pmatrix}$ in S_3.

   b. Find the orders of the permutations
   $\begin{pmatrix} 1 & 2 & 3 & 4 \\ 1 & 3 & 2 & 4 \end{pmatrix}$, $\begin{pmatrix} 1 & 2 & 3 & 4 \\ 2 & 3 & 1 & 4 \end{pmatrix}$, and $\begin{pmatrix} 1 & 2 & 3 & 4 \\ 3 & 4 & 1 & 2 \end{pmatrix}$ in S_4.

4. In S_4 find $\left\langle \begin{pmatrix} 1 & 2 & 3 & 4 \\ 1 & 3 & 4 & 2 \end{pmatrix} \right\rangle$, the subgroup of S_4 generated by $\begin{pmatrix} 1 & 2 & 3 & 4 \\ 1 & 3 & 4 & 2 \end{pmatrix}$.

5. Consider the function on X_n represented by $\begin{pmatrix} 1 & 2 & 3 & \cdots & n \\ y_1 & y_2 & y_3 & \cdots & y_n \end{pmatrix}$.

Permutation groups

Under what conditions on the y_i's, $i = 1, 2, \ldots, n$, is this function a member of S_n?

We will develop another notation for a permutation that will give us more information about the function. We will study the decomposition of a permutation into its disjoint cycles.

EXAMPLE 9.4

If we look carefully at the permutation

$$s = \begin{pmatrix} 1 & 2 & 3 & 4 & 5 & 6 & 7 & 8 \\ 2 & 3 & 1 & 8 & 4 & 7 & 6 & 5 \end{pmatrix}$$

we see that $s(1) = 2$, $s(2) = 3$, $s(3) = 1$. The subset $\{1, 2, 3\}$ is treated as a unit by s. Similarly $s(4) = 8$, $s(8) = 5$, $s(5) = 4$ so that the subset $\{4, 8, 5\}$ is treated separately by s. $s(6) = 7$ and $s(7) = 6$, so $\{6, 7\}$ is the final subset of this kind. That is, the permutation s induces a partition $\{\{1, 2, 3\} \{4, 8, 5\} \{6, 7\}\}$ on the set X_n.

We will pursue this thought further and show that every permutation is associated with a partition of X_n. We will use this partition as a basis for a new notation for the permutation.

DEFINITION

Let s be any permutation in S_n and let x be any element of X_n. By the **orbit of x under s** we mean the set $\{s^n(x) | n \in Z\}$.

EXAMPLE 9.5

In the permutation $\begin{pmatrix} 1 & 2 & 3 & 4 & 5 & 6 \\ 3 & 1 & 4 & 2 & 6 & 5 \end{pmatrix} \in S_6$, we will find the orbit of the integer $1 \in X_n$:

$s^0(1) = 1$
$s^1(1) = 3$
$s^2(1) = s(3) = 4$
$s^3(1) = s(4) = 2$
$s^4(1) = s(2) = 1$.

Considering higher powers of s would give us a repetition of the elements obtained. The orbit of 1 is $\{1, 3, 4, 2\}$. $\{1, 3, 4, 2\}$ is also the orbit of the integers 3, 4, and 2 while the orbit of $5 = \{5, 6\} = $ the orbit of 6.

For an element $s \in S_n$, we define a relation on the set X_n by: x is related to y means

"x belongs to the orbit of y under s."

THEOREM 9.4

The relation "x belongs to the orbit of y under s" is an equivalence relation.

The proof is left as an exercise.

Note that Theorem 9.4 guarantees that every permutation in S_n is associated with a partition of X_n. By writing each set of the partition in a specific order, we get a representation for the permutation.

DEFINITION

Let $s \in S_n$. A **cycle** of s is of the form $(x, s(x), s^2(x), \ldots, s^{k-1}(x))$ for any $x \in X_n$ where $s^k(x) = x$ and k is the smallest positive integer for which this is true.

Note that in making this definition we are assuming that there is a positive integer k such that $s^k(x) = x$. You will be asked to prove this in the exercises. You will also be asked to prove that all entries in a cycle of s are distinct.

Also note that any cycle of $s \in S_n$ is a set of the partition of X_n, induced by s, written in a fixed order. Therefore any two cycles of s either contain the same entries or are disjoint.

EXAMPLE 9.6

For $s = \begin{pmatrix} 1 & 2 & 3 & 4 & 5 & 6 & 7 & 8 \\ 2 & 3 & 1 & 8 & 4 & 7 & 6 & 5 \end{pmatrix} \in S_8$, decompose s into disjoint cycles. Starting with the element $1 \in X_8$,

$s(1) = 2, \qquad s^2(1) = s(2) = 3, \qquad s^3(1) = s(3) = 1$

so that one cycle of s is (1 2 3).

Starting again with an element not in the first cycle, we choose 4:

$4 \to 8 \to 5 \to 4$

so that the next cycle is (4 8 5).

Since $6 \to 7 \to 6$, the final cycle of s is (6 7).

We can represent s as a product of its disjoint cycles as $s = $ (1 2 3)(4 8 5)(6 7).

Note that s could also have been written as

$s = $ (2 3 1)(5 4 8)(7 6)

since we can start anywhere within a cycle. However, the circular order within each cycle is fixed; that is, (1 3 2) is not a cycle of s since, if it were,

$s(1) = 3, \qquad s(3) = 2, \qquad \text{and} \qquad s(2) = 1$

which contradicts the definition of s.

Permutation groups

We can use the cyclic notation to compose two permutations.

EXAMPLE 9.7

Suppose $s, t \in S_4$, $s = \begin{pmatrix} 1 & 2 & 3 & 4 \\ 3 & 2 & 4 & 1 \end{pmatrix}$, and $t = \begin{pmatrix} 1 & 2 & 3 & 4 \\ 1 & 4 & 2 & 3 \end{pmatrix}$. In cyclic notation, $s = (1\ 3\ 4)(2)$ and $t = (1)(2\ 4\ 3)$. We write $s \circ t$ as

$$s \circ t = (1\ 3\ 4)(2)(1)(2\ 4\ 3)$$

and follow each element through from right to left:

1 is unchanged by the cycle (2 4 3)

1 → 1 in the cycle (1)

1 is unchanged in the cycle (2)

and

1 → 3 in the cycle (1 3 4).

Therefore $(s \circ t)(1) = 3$ and $(1\ 3\ \ldots)$ begins a cycle of $s \circ t$. We continue:

3 → 2 in the cycle (2 4 3)

2 is unchanged by the cycle (1)

2 → 2 in (2)

and

2 is unchanged in (1 3 4).

Therefore $(s \circ t)(3) = 2$ and $(1\ 3\ 2\ \ldots)$ starts the cycle of $s \circ t$.

2 → 4 in (2 4 3)

4 is unchanged in cycles (1) and (2)

and

4 → 1 in (1 3 4).

Therefore $(s \circ t)(2) = 1$ and $(1\ 3\ 2)$ is a complete cycle of $s \circ t$. $(s \circ t)$ must map 4 to 4 so that

$$s \circ t = (1\ 3\ 2)(4).$$

A final note here: Within S_n, for a fixed value of n, a simplified notation is usually achieved by not writing down all single element cycles. They do not add any information. That is, (2 3 4) is the same as (1)(2 3 4) if the permutation is an element of S_4.

EXERCISES

1. Decompose each of the following permutations into its disjoint cycle representation. Use the last example as a guide.

   a. $\begin{pmatrix} 1 & 2 & 3 & 4 \\ 2 & 1 & 4 & 3 \end{pmatrix}$

   b. $\begin{pmatrix} 1 & 2 & 3 & 4 \\ 2 & 1 & 3 & 4 \end{pmatrix}$

   c. $\begin{pmatrix} 1 & 2 & 3 & 4 & 5 & 6 & 7 & 8 \\ 3 & 4 & 1 & 5 & 2 & 7 & 8 & 6 \end{pmatrix}$

   d. $\begin{pmatrix} 1 & 2 & 3 & 4 & 5 & 6 & 7 & 8 \\ 4 & 7 & 5 & 3 & 8 & 2 & 1 & 6 \end{pmatrix}.$

2. Find the orders of the above permutations.

3. The following permutations are written in cyclic form. Rewrite them in standard form.
   a. (1 4 2)(3) in S_4.
   b. (1 2 3) in S_5.
   c. (5 4 8 3 2)(1 7) in S_8.
   d. (1 4 3 2)(9 5 6) in S_9.

4. What is the identity of S_n in cyclic notation?

5. What is $(x_1 x_2 \cdots x_r)^{-1}$ in cyclic notation?

6. Prove Theorem 9.4.

7. a. Prove that for every $x \in X_n$ and $s \in S_n$, there is a positive integer m such that $s^m(x) = x$.
   b. Prove that $m \leq n$.

8. If $(x, s(x), s^2(x), \ldots, s^{m-1}(x))$ is a cycle of $s \in S_n$, prove that for all integers j and k, $1 \leq j, k, \leq m$, $s^j(x) \neq s^k(x)$ if $j \neq k$.

9. Show that disjoint cycles commute.

10. a. If s consists of a product of two disjoint cycles of length p and q where p and q are distinct primes, show that the order of s is pq.
    b. If, for any positive integers k and m, s is a product of a cycle of length k and a cycle of length m, what is the order of s?

11. Perform the following compositions.
    a. (7 6 4 5)(1 4)(2 3 8)(1 7 6)(2 4)
    b. (1 3 2)(4 5)(3 1 2)(4 5)
    c. (1 2)(1 3)(1 4)(1 5)(1 6)

12. Let s be a permutation with a given cycle structure. That is, if $s \in S_n$, then s has c_1 cycles of length 1, c_2 cycles of length 2, c_3 cycles of length 3, ..., c_n cycles of length n. Show that $1c_1 + 2c_2 + \cdots + nc_n = n$.

13. Denote the cycle structure of $s \in S_n$ by

    $\langle c_1, c_2, \ldots, c_n \rangle.$

    Let s be a permutation with cycle structure $\langle c_1, c_2, \ldots, c_n \rangle$. Let t be any other permutation in S_n. Prove that $t^{-1} \circ s \circ t$ has the same cycle structure as s.

Permutation groups

We turn now to a very important notion in our study of permutation groups.

DEFINITION
A **transposition** is a cycle of length 2.
For example, (x_1, x_2) is a transposition.
Note that any cycle can be written as a product of transpositions since

$$(x_1, x_2, \ldots, x_m) = (x_1, x_m)(x_1, x_{m-1}) \cdots (x_1 x_3)(x_1 x_2).$$

(You are asked to verify this in the exercises.)

EXAMPLE 9.8
If $s = (1\ 6\ 5)(2\ 4\ 3\ 7)$, then $(1\ 6\ 5) = (1\ 5)(1\ 6)$ and $(2\ 4\ 3\ 7) = (2\ 7)(2\ 3)(2\ 4)$ so that $s = (1\ 6\ 5)(2\ 4\ 3\ 7) = (1\ 6)(1\ 5)(2\ 7)(2\ 3)(2\ 4)$.

Note that there are many different ways of writing a permutation as a product of transpositions. For instance,

$(1\ 3\ 2) = (1\ 2)(1\ 3)$
$ = (2\ 1)(2\ 1)(1\ 2)(1\ 3)(2\ 3)(3\ 2)$
$ = (3\ 1)(3\ 2).$

Note also that the decomposition of a permutation into transpositions destroys the disjointness of the representation. For example, $(1\ 2\ 3) = (1\ 3)(1\ 2)$ and the integer 1 belongs to both transpositions.

However, the next theorem gives a very useful invariant of the decomposition of a permutation into transpositions. We include the proof for completeness; however, the proof leads to no increased understanding of the theorem and could be omitted.

THEOREM 9.5
Let s be a permutation. If s has one representation as a product of an even number of transpositions, then every representation of s as a product of transpositions also has an even number.

Proof: Let $s \in S_n$. Consider the expression

$$p = (x_1 - x_2)(x_1 - x_3) \cdots (x_1 - x_n)(x_2 - x_3)(x_2 - x_4) \cdots (x_2 - x_n)$$
$$\cdot (x_3 - x_4) \cdots (x_3 - x_n) \cdots (x_{n-1} - x_n).$$

That is, p is a product of all the terms $(x_i - x_j)$ where $i < j$. We define the action of s on p by

$$s(p) = (x_{s(1)} - x_{s(2)})(x_{s(1)} - x_{s(3)}) \cdots (x_{s(n-1)} - x_{s(n)}).$$

Note that $s(p)$ is uniquely determined by the permutation s.

We will verify the following statements:
1. $(s \circ t)(p) = s(t(p))$ for any permutations s and t.
2. t is a transposition $t(p) = -p$.

1. $(s \circ t)(p) =$ the product of all terms $(x_{(s \circ t)(i)} - x_{(s \circ t)(j)})$, where $1 \le i < j \le n$. But $t(p) =$ the product of all terms $(x_{t(i)} - x_{t(j)})$ for $i < j$ and $s(t(p))$ is the product of all terms

$$(x_{s(t(i))} - x_{s(t(j))}) = (x_{(s \circ t)(i)} - x_{(s \circ t)(j)}).$$

2. Let $t = (k\ l)$ be a transposition with $k < l$. Recall that p consists of all products $(x_i - x_j)$ for $i < j$. Many of these terms will not involve k or l and they will not be changed in the expression $t(p)$. For the other terms, we pair them off as follows:

a. For $m < k < l$, pair off the factors $(x_m - x_k)$ and $(x_m - x_l)$. In $t(p)$, this pairs off $(x_m - x_l)$ and $(x_m - x_k)$; the products are the same:

$$(x_m - x_k)(x_m - x_l) = (x_{t(m)} - x_{t(k)})(x_{t(m)} - x_{t(l)}).$$

b. For $m > l > k$, pair off the factors $(x_k - x_m)$ and $(x_l - x_m)$ in p. Again this pairs off $(x_l - x_m)$ and $(x_k - x_m)$ in $t(p)$; and again the product in $t(p)$ equals that in p.

c. For $k < m < l$, we pair off $(x_k - x_m)$ and $(x_m - x_l)$ in p. The product $(x_k - x_m)(x_m - x_l)$ is mapped to $(x_l - x_m)(x_m - x_k)$ in $t(p)$ and again the products are equal.

Finally $(x_k - x_l)$ in p is mapped to $(x_l - x_k) = -(x_k - x_l)$ in $t(p)$. At last, $t(p) = -p$.

Now we use facts 1 and 2 to complete the proof of the theorem.

Assume s has one representation as a product of an even number of transpositions

$$s = t_1 \circ t_2 \circ \cdots \circ t_{2m}.$$

Then

$$s(p) = t_1(t_2(t_3 \cdots (t_{2m}(p)) \cdots))$$
$$= p$$

because the sign of p has been changed an even number of times. Therefore any other representation of s as a product of transpositions must contain an even number of factors also. ∎

Note that Theorem 9.5 also implies that if one representation of a permutation is as the product of an odd number of transpositions, then every other representation as a product of transpositions also contains an odd number of factors.

Because of Theorem 9.5 we can make the following definitions.

DEFINITION

A permutation is **even** if it can be written as the product of an even number of transpositions. A permutation of S_n is **odd** if it is the product of an odd number of transpositions.

The following results are immediate.

LEMMA

1. The product of two even permutations is even.
2. The product of two odd permutations is even.
3. The product of an even permutation and an odd permutation is odd.

The proof is left as an exercise.

DEFINITION

For $n \geq 2$, let A_n be the subset of S_n consisting of all even permutations.

THEOREM 9.6

A_n is a subgroup of S_n.

Proof: Note that $A_n \neq \emptyset$ since the identity permutation can be written as (1 2)(2 1) and is an even permutation.

Choose any elements x and y in A_n. The permutation y^{-1} is even since y is even and the product $y \circ y^{-1}$ is the identity which is even.

Also, $x \circ y$ is an even permutation by the last lemma. By Theorem 5.1, $A_n < S_n$. ∎

THEOREM 9.7

$A_n \triangleleft S_n$ for $n \geq 2$.

Proof: Let $G = \{0, 1\}$ be the cyclic group of order 2 under addition modulo 2. Define a function $\alpha: S_n \to G$ by

$$\alpha(s) = \begin{cases} 0, & \text{if } s \in A_n \\ 1, & \text{if } s \notin A_n. \end{cases}$$

We will show that α is a homomorphism with kernel A_n. Consider two permutations $s, t \in S_n$.

Case 1: If s and t are even, then $\alpha(s \circ t) = 0$ since $s \circ t$ is even. $\alpha(s) + \alpha(t) = 0 + 0 = 0$ also.

Case 2: If s is even and t is odd, then $s \circ t$ is odd so that $\alpha(s) + \alpha(t) = 0 + 1 = 1 = \alpha(s \circ t)$.

The cases where s is odd and t is even and where both s and t are odd can be proved similarly.

By the definition of α, $A_n = \ker(\alpha)$. Therefore $A_n \triangleleft S_n$ since it is the kernel of a homomorphism defined on S_n. ∎

COROLLARY

$$|A_n| = \frac{n!}{2}.$$

Proof: Since $\alpha : S_n \to \{0, 1\}$ is a homomorphism with kernel A_n, then $S_n | A_n$ is isomorphic to $\{0, 1\}$. Therefore

$$|S_n | A_n| = 2$$

and

$$|A_n| = \frac{|S_n|}{|S_n | A_n|} = \frac{n!}{2}. \quad \blacksquare$$

EXAMPLE 9.9

Let us look at some particular groups S_n and A_n with their permutations written in cyclic notation. Let e be the identity of each S_n.

  a. For $n = 2$, $S_2 = \{e, (1\ 2)\}$ and $A_2 = \{e\}$.
  b. If $n = 3$, $S_3 = \{e, (1\ 2), (1\ 3), (2\ 3), (1\ 2\ 3), (1\ 3\ 2)\}$ while $A_3 = \{e, (1\ 2\ 3), (1\ 3\ 2)\}$. Here $|A_3| = 3$.
  c. When $n = 4$, $|S_4| = 24$ while $|A_4| = 12$.

The orders of A_2 and A_3 do not allow any nontrivial normal subgroups. A_4 does have normal subgroups. The subgroup of S_4 that is isomorphic to the Klein four-group (by Cayley's Theorem) is normal in A_4.

Also, A_4 is an example of a group of order 12 that has no subgroup of order 6. This is proved by showing that any subgroup of A_4 of order 6 cannot contain an element of order 2. This contradicts Cauchy's Theorem. The proof depends on the specific operation in A_4 and is not necessarily true for other groups of order 12.

EXERCISES

1. a. Prove that the product of either two even or two odd permutations is an even permutation.
   b. Prove that the product of an even permutation and an odd permutation is an odd permutation.
2. Show that A_n is not abelian for $n \geq 4$.

3. Which of the following permutations are even?
   a. (4 5 6)
   b. (1 2)(7 8 9 10)
   c. (1 2)(1 3)(1 4)(1 5)(1 6)
   d. (1 2 3)(4 5 6)(7 8 9)
4. Verify that $(x_1 x_2 \cdots x_n) = (x_1 x_n)(x_1 x_{n-1}) \cdots (x_1 x_3)(x_1 x_2)$.
5. Prove that every cycle $(i\ j\ k)$, consisting of 3 entries, is even.
6. Find all subgroups of A_4.
7. Let $s \in S_n$ be an element of order 2. Prove that the subgroup $H = \{e, s\}$ is not normal in S_n if $n > 2$.
8. Prove that if H is a subgroup of S_n, then either $H < A_n$ or exactly half of the elements of H are even.

Our next major result is that for $n \geq 5$, A_n has no normal subgroups other than itself and $\{e\}$. The proof proceeds with a series of lemmas and depends heavily on the fact that every 3-cycle $\in A_n$.

LEMMA 1

Let n be a positive integer greater than or equal to 5 and let A_n be the alternating group. Let H be a subgroup of A_n that contains all 3-cycles. Then $H = A_n$.

Proof: $H \subseteq A_n$ by definition. We will show $A_n \subseteq H$. Let $s \in A_n$ be any even permutation; then

$$s = (t_1 \circ t_2) \circ (t_3 \circ t_4) \circ \cdots \circ (t_{2m-1} \circ t_{2m})$$

a product of m pairs of transpositions. We will show that each pair of transpositions, $t_i \circ t_j$, is the identity or can be written as a product of 3-cycles by considering the cases (1) $t_i \circ t_j = e$, (2) $t_i \circ t_j = (x\ y) \circ (x\ z)$ for integers x, y, and z, and (3) $t_i \circ t_j = (x\ y)(z\ w)$ for distinct integers x, y, z, and w, separately.
  1. Since $H < A_n$, $e \in H$.
  2. $(x\ y) \circ (x\ z) = (x\ z\ y) \in H$ since it is a 3-cycle.
  3. $(x\ y)(z\ w) = (x\ z\ y)(x\ z\ w) \in H$.

Every pair of transpositions in s can be written as a product of 3-cycles and therefore s can be written as a product of 3-cycles and $s \in H$.
Therefore $A_n \subseteq H$, which completes the proof. ∎

LEMMA 2

Let H be a normal subgroup of A_n ($n \geq 5$). Assume H contains one 3-cycle. Then $H = A_n$.

Proof: Assume that $(x_1, x_2, x_3) \in H$. We wish to show that every 3-cycle is in H. Since $H \triangleleft A_n$, for any $g \in A_n$ and $h \in H$, $g^{-1} \circ h \circ g \in H$.

Let $g = (x_1\ x_4\ x_2)$ and $h = (x_1\ x_2\ x_3)$. Then
$$g^{-1} \circ h \circ g = (x_2\ x_4\ x_1)(x_1\ x_2\ x_3)(x_1\ x_4\ x_2)$$
$$= (x_1)(x_2\ x_4\ x_3)$$
$$= (x_2\ x_4\ x_3).$$

Thus it is possible to change one element of $(x_1\ x_2\ x_3)$ and get a 3-cycle which must be in H. By changing one element at a time, it is possible to show that every 3-cycle must be in H. (Why?) Therefore $H = A_n$ by the previous lemma. ∎

LEMMA 3
Let H be a normal subgroup of A_n other than $\{e\}$. Then H contains a 3-cycle.

Proof: Assume each element of H is written as a product of disjoint cycles.

1. If some permutation $s \in H$ has a cycle of length ≥ 4, we write
$$s = (x_1\ x_2\ x_3\ \cdots\ x_r)(\ )\cdots(\ ).$$
Let $t = (x_2, x_3, x_4) \in A_n$. Since $H \triangleleft A_n$, $t^{-1} \circ s \circ t \in H$.
$$t^{-1} \circ s \circ t = (x_4\ x_3\ x_2)(x_1\ x_2\ \cdots\ x_r)(\ )\cdots(\ )(x_2\ x_3\ x_4)$$
$$= (x_1\ x_4\ x_2\ x_3\ x_5\ \cdots\ x_r)(\ )\cdots(\ ).$$
(Follow the subscripts to ease the computations.)
Note that the cycles other than $(x_1, x_2, \ldots x_r)$ are unaffected by this product since they are disjoint from all other factors.

Finally $(t^{-1} \circ s \circ t) \circ s^{-1} \in H$ since H is closed under $\circ$, but
$$(t^{-1} \circ s \circ t) \circ s^{-1} = (x_1\ x_4\ x_2\ x_3\ x_5\ \cdots\ x_r)(\ )\cdots(\ )$$
$$\circ (x_r x_{r-1} \cdots x_1)(\ )^{-1} \cdots (\ )^{-1}$$
$$= (x_1)(x_2\ x_4\ x_5)(x_3)$$
$$= (x_2\ x_4\ x_5) \quad \text{which is a 3-cycle in } H.$$

2. We will show that if a permutation in H has two or more of its disjoint cycles of length 3, then H has a permutation with a cycle of length ≥ 4. Let $s = (x_1\ x_2\ x_3)(x_4\ x_5\ x_6)(\ )\cdots(\ ) \in H$ and let $t = (x_3\ x_4\ x_5) \in A_n$. We perform the same products as in part 1.
$$t^{-1} \circ s \circ t = (x_1\ x_2\ x_5)(x_3\ x_4\ x_6)(\ )\cdots(\ ) \in H.$$
Finally $(t^{-1} \circ s \circ t) \circ s^{-1} = (x_1\ x_6\ x_3\ x_4\ x_5) \in H$. Therefore H satisfies the hypothesis of part 1 and has a 3-cycle.

3. If $s \in H$ has one cycle whose length is 3 and all its other cycles are of length 1 or 2, then $s \circ s \in H$ and is a 3-cycle. (Why?)

4. If every s in H has all of its cycles of length 1 or 2, then consider the permutation

$$s = (x_1\ x_2)(x_3\ x_4)(\quad)\cdots(\quad)$$

in H. Now, since we have assumed that $n \geq 5$, the permutation $t = (x_1\ x_2\ x_5) \in A_n$. For the same reasoning as before, $(t^{-1} \circ s \circ t) \circ s^{-1} \in H$, but

$$(t^{-1} \circ s \circ t) \circ s^{-1} = (x_1\ x_2\ x_5)(x_3)(x_4)$$
$$= (x_1\ x_2\ x_5), \quad \text{a 3-cycle in } H. \blacksquare$$

THEOREM 9.8

If $n \geq 5$, the only normal subgroups of A_n are A_n itself and $\{e\}$.

Proof: Assume $H \triangleleft A_n$, $H \neq \{e\}$. By Lemma 3, there is a 3-cycle in H. Lemma 2 then implies that $H = A_n$. $\blacksquare$

This property of the alternating group leads to the following definitions.

DEFINITIONS

1. A group G is called **simple** if it has no normal subgroups other than G and $\{e\}$.

2. Let G be any group. A sequence of normal subgroups

$$\{e\} \triangleleft N_1 \triangleleft N_2 \triangleleft \cdots \triangleleft N_k \triangleleft G$$

is called a **normal series**. (Note that we do not assume that each $N_i \triangleleft G$, only that $N_i \triangleleft N_{i+1}$.)

3. If there is a subgroup M with $N_i \triangleleft M \triangleleft N_{i+1}$, then the normal series

$$\{e, N_1, N_2, \ldots, N_i, M, N_{i+1}, \ldots, N_k, G\}$$

is a **refinement** of the normal series

$$\{e, N_1, N_2, \ldots, N_k, G\}.$$

4. If $\{e, N_1, \ldots, N_k, G\}$ is a normal series for which no refinement is possible, then $\{e, N_1, \ldots, N_k, G\}$ is called a **composition series** for G.

5. If $\{e, N_1, N_2, \ldots, N_k, G\}$ is a composition series for G, then the factor groups

$$G|N_k, N_k|N_{k-1}, \ldots, N_{i+1}|N_i, \ldots, N_1$$

are called **composition factors** for G.

6. If there is a composition series for G whose composition factors are all abelian, we say that G is a **solvable** group.

We use the concept of solvable groups and the next theorem in our study of fields.

THEOREM 9.9
S_n is not solvable if $n \geq 5$.

Proof: $A_n \triangleleft S_n$ so that S_n is not simple. However, A_n is simple by Theorem 9.8. Therefore the series

$$\{e\} \triangleleft A_n \triangleleft S_n$$

has no refinement. But $A_n | \{e\}$ is not abelian.

Assume $\{e\} \triangleleft H \triangleleft S_n$ is another normal series of S_n. Then $H \cap A_n \triangleleft A_n$ (why?) and $H \cap A_n = A_n$ or $\{e\}$ by Theorem 9.8.

1. If $H \cap A_n = A_n$, then the series is the same as the one above.
2. If $H \cap A_n = \{e\}$, then $H - \{e\}$ consists only of odd permutations. Since the product of any two odd permutations is even, $H = \{e\}$ or $\{e, s\}$ where the order of s is 2. But no two element subgroup of S_n is normal in S_n.

Therefore there does not exist a composition series of S_n with abelian composition factors and S_n is not solvable. ∎

EXERCISES
1. Write out composition series for S_2, S_3, S_4, S_5. Which are solvable?
2. Show that every finite abelian group is solvable.
3. Show that if G is a finite solvable group, then its composition factors are all cyclic of prime order.
4. Let G be a finite simple group of order n, n not prime. Show that G is not solvable.
5. Not every infinite group has a composition series. Show that $(Z, +)$ has none.

RING THEORY

10
Examples and axioms

In studying groups we saw that a group is a set with an operation on it which satisfies certain rules, called axioms. Generally, studying algebra consists of studying sets, with one or more operations defined on them, which satisfy certain axioms. As with groups, the definition of a ring will make more sense if we investigate some examples first. In this chapter we will look at some examples of rings before we proceed to the formal definition.

EXAMPLE 10.1

Let us consider the real numbers, $\mathbb{R}$. The set $\mathbb{R}$ has two operations defined on it: addition and multiplication. Recall that $(\mathbb{R}, +)$ is an abelian group and $(\mathbb{R} - \{0\}, \cdot)$ is also a group. In particular, $\mathbb{R}$ is closed under multiplication (that is, if x and y are in $\mathbb{R}$, then $x \cdot y \in \mathbb{R}$), and multiplication is associative ($x \cdot (y \cdot z) = (x \cdot y) \cdot z$ for any $x, y, z \in \mathbb{R}$). (We will not need to consider the other properties of $\mathbb{R}$ under multiplication to study $\mathbb{R}$ as a ring.) Finally, $\mathbb{R}$ satisfies both distributive laws: for all $x, y, z \in \mathbb{R}$,

1. $x(y + z) = xy + xz$ and
2. $(y + z)x = yx + zx$.

We will look at a few more examples until a pattern becomes clear. Some of the examples are interesting in themselves.

EXAMPLE 10.2

Our next example is the even integers, E. As above, $(E, +)$ is an abelian group. Also, if x and y are even integers, then $x \cdot y$ is also. Therefore E is closed under multiplication. Finally, multiplication of integers is associative and the distributive laws hold.

EXAMPLE 10.3
Consider $M_{2 \times 2}$, the set of all 2×2 matrices with real entries and the operations of matrix addition and matrix multiplication. First, we know that $M_{2 \times 2}$ is an abelian group. Second, if M and N are both 2×2 matrices, then $M \cdot N$ is also a 2×2 matrix; that is, the set $M_{2 \times 2}$ is closed under multiplication. We also need to verify that (1) the associative law holds for matrix multiplication and that (2) both distributive laws hold.

1. Does $X(Y \cdot Z) = (X \cdot Y)Z$? Let $X = \begin{pmatrix} a & b \\ c & d \end{pmatrix}$, $Y = \begin{pmatrix} e & f \\ g & h \end{pmatrix}$, and $Z = \begin{pmatrix} i & j \\ k & l \end{pmatrix}$. Then

$$X(Y \cdot Z) = \begin{pmatrix} a & b \\ c & d \end{pmatrix} \left[\begin{pmatrix} e & f \\ g & h \end{pmatrix} \cdot \begin{pmatrix} i & j \\ k & l \end{pmatrix} \right]$$

$$= \begin{pmatrix} a & b \\ c & d \end{pmatrix} \begin{pmatrix} ei+fk & ej+fl \\ gi+hk & gj+hl \end{pmatrix}$$

$$= \begin{pmatrix} aei+afk+bgi+bhk & aej+afl+bgj+bhl \\ cei+cfk+dgi+dhk & cej+cfl+dgj+dhl \end{pmatrix}$$

while

$$(X \cdot Y)Z = \left[\begin{pmatrix} a & b \\ c & d \end{pmatrix} \cdot \begin{pmatrix} e & f \\ g & h \end{pmatrix} \right] \cdot \begin{pmatrix} i & j \\ k & l \end{pmatrix}$$

$$= \begin{pmatrix} ae+bg & af+bh \\ ce+dg & cf+dh \end{pmatrix} \cdot \begin{pmatrix} i & j \\ k & l \end{pmatrix}$$

$$= \begin{pmatrix} aei+bgi+afk+bhk & aej+bgj+afl+bhl \\ cei+dgi+cfk+dhk & cej+dgj+cfl+dhl \end{pmatrix}.$$

Since both products are equal, the associative law holds.

We will show that one distributive law holds; that is, $X(Y + Z) = XY + XZ$ for all X, Y, Z in $M_{2 \times 2}$. Again, let $X = \begin{pmatrix} a & b \\ c & d \end{pmatrix}$, $Y = \begin{pmatrix} e & f \\ g & h \end{pmatrix}$, and $Z = \begin{pmatrix} i & j \\ k & l \end{pmatrix}$.

$$X(Y + Z) = \begin{pmatrix} a & b \\ c & d \end{pmatrix} \left[\begin{pmatrix} e & f \\ g & h \end{pmatrix} + \begin{pmatrix} i & j \\ k & l \end{pmatrix} \right]$$

$$= \begin{pmatrix} a & b \\ c & d \end{pmatrix} \begin{pmatrix} e+i & f+j \\ g+k & h+l \end{pmatrix}$$

$$= \begin{pmatrix} ae+ai+bg+bk & af+aj+bh+bl \\ ce+ci+dg+dk & cf+cj+dh+dl \end{pmatrix}.$$

Examples and axioms

$$XY + XZ = \begin{pmatrix} a & b \\ c & d \end{pmatrix}\begin{pmatrix} e & f \\ g & h \end{pmatrix} + \begin{pmatrix} a & b \\ c & d \end{pmatrix}\begin{pmatrix} i & j \\ k & l \end{pmatrix}$$

$$= \begin{pmatrix} ae + bg & af + bh \\ ce + dg & cf + dh \end{pmatrix} + \begin{pmatrix} ai + bk & aj + bl \\ ci + dk & cj + dl \end{pmatrix}$$

$$= \begin{pmatrix} ae + bg + ai + bk & af + bh + aj + bl \\ ce + dg + ci + dk & cf + dh + cj + dl \end{pmatrix}.$$

Since the results are equal, this distributive law holds. The fact that $(Y + Z)X = YX + ZX$ is left as an exercise.

EXAMPLE 10.4

In this example, we begin with a group $Z_2 = \{0, 1\}$ of order 2 under addition modulo 2. Z_2 is cyclic and therefore abelian. $(Z_2, +)$ has the group table

+	0	1
0	0	1
1	1	0

We define another operation $\cdot$ on Z_2 by

$\cdot$	0	1
0	0	0
1	0	1

By the definition of the operation, for any x and y in Z_2, $x \cdot y$ is also in Z_2. We need to verify the associative and distributive laws.

To verify the associative law, we need to consider the following eight products:

$0 \cdot (0 \cdot 0) = 0 = (0 \cdot 0) \cdot 0 \qquad 1 \cdot (0 \cdot 0) = 0 = (1 \cdot 0) \cdot 0$
$0 \cdot (0 \cdot 1) = 0 = (0 \cdot 0) \cdot 1 \qquad 1 \cdot (0 \cdot 1) = 0 = (1 \cdot 0) \cdot 1$
$0 \cdot (1 \cdot 0) = 0 = (0 \cdot 1) \cdot 0 \qquad 1 \cdot (1 \cdot 0) = 0 = (1 \cdot 1) \cdot 0$
$0 \cdot (1 \cdot 1) = 0 = (0 \cdot 1) \cdot 1 \qquad 1 \cdot (1 \cdot 1) = 1 = (1 \cdot 1) \cdot 1.$

To show that $x(y + z) = xy + xz$ for all x, y, z in Z_2, we consider the following eight cases.

$0 \cdot (0 + 0) = 0 = 0 \cdot 0 + 0 \cdot 0 \qquad 1 \cdot (0 + 0) = 0 = 1 \cdot 0 + 1 \cdot 0$
$0 \cdot (0 + 1) = 0 = 0 \cdot 0 + 0 \cdot 1 \qquad 1 \cdot (0 + 1) = 1 = 1 \cdot 0 + 1 \cdot 1$
$0 \cdot (1 + 0) = 0 = 0 \cdot 1 + 0 \cdot 0 \qquad 1 \cdot (1 + 0) = 1 = 1 \cdot 1 + 1 \cdot 0$
$0 \cdot (1 + 1) = 0 = 0 \cdot 1 + 0 \cdot 1 \qquad 1 \cdot (1 + 1) = 0 = 1 \cdot 1 + 1 \cdot 1.$

If we note that the product · defined on Z_2 is commutative (the table is symmetric), we can write

$$(y + z)x = x(y + z) = xy + xz = yx + zx$$

verifying the second distributive law.

Example 10.4 is a special case of a more general class of rings arising out of addition and multiplication modulo n.

EXAMPLE 10.5

In Chapter 6, when we discussed equivalence relations, we defined a relation on the integers by

$$a \equiv b \pmod{n} \quad \text{or} \quad a \equiv_n b$$

read "a is congruent to b mod n" and meaning $n|(a - b)$. We prove that congruence modulo n is an equivalence relation with equivalence classes

$$[0] = \{kn | k \in Z\} = [n] = [2n] = \cdots$$
$$[1] = \{kn + 1 | k \in Z\} = [n + 1] = [2n + 1] = \cdots$$
$$\vdots$$
$$[n] = \{kn + (n - 1) | k \in Z\} = [n + (n - 1)] = [2n + (n - 1)] = \cdots.$$

The equivalence classes form a partition of Z; that is, every integer belongs to exactly one equivalence class. We defined an operation + on the set Z_n of equivalence classes by

$$[a] + [b] = [a + b].$$

We proved that the operation + is well defined and that the system $(Z_n, +)$ is a group. We also noted that the group is essentially the same as the group $(G_n, \oplus)$ defined in Chapter 3.

We defined another operation on the set Z_n of equivalence classes by

$$[a] \cdot [b] = [ab].$$

To complete a proof that the set $Z_n = \{[0], [1], \ldots, [n - 1]\}$ is a ring with the operations + and ·, we need to show that (1) $(Z_n, +)$ is abelian, (2) the operation · is well defined, (3) · is associative, and (4) the distributive laws are satisfied.

1. To show that Z_n is abelian under the operation +, we note that for any $[i]$ and $[j]$ in Z_n,

$$[i] + [j] = [i + j] = [j + i] = [j] + [i].$$

This verification depends on the fact that addition of integers is commutative.

2. To show that the operation · is well defined, we need to show that the product $[a] \cdot [b] = [ab]$ does not depend on the representatives of the equivalence classes that are used for the computation.

Choose any elements $c \in [a]$ and $d \in [b]$. This implies $c = a + kn$ and $d = b + ln$ for some integers k and l. Therefore

$$\begin{aligned}[c][d] = [cd] &= [(a+kn)(b+ln)] \\ &= [ab + knb + lna + lkn^2] \\ &= [ab + n(kb + la + lkn)] \\ &= [ab] \quad \text{(why?)}.\end{aligned}$$

Therefore the operation · is well defined.

3. To show that · is associative, choose $[a], [b], [c] \in Z_n$,

$$\begin{aligned}([a] \cdot [b]) \cdot [c] &= [ab] \cdot [c] \\ &= [(ab)c] \\ &= [a(bc)] \\ &= [a] \cdot [bc] = [a] \cdot ([b] \cdot [c]).\end{aligned}$$

The proof depends on the associativity of multiplication of integers.

4. To prove the distributive laws, choose any elements $[a], [b], [c]$ in Z_n; then

$$\begin{aligned}[a]([b] + [c]) &= [a][b + c] \\ &= [a(b + c)] \\ &= [ab + ac] \\ &= [ab] + [ac] \\ &= [a][b] + [a][c].\end{aligned}$$

Again the proof depends on the distributive property of the integers. The second distributive law is proved similarly. We conclude this example by giving addition and multiplication tables for Z_6. (The tables for Z_5 are given in Chapter 6.)

+	[0]	[1]	[2]	[3]	[4]	[5]
[0]	[0]	[1]	[2]	[3]	[4]	[5]
[1]	[1]	[2]	[3]	[4]	[5]	[0]
[2]	[2]	[3]	[4]	[5]	[0]	[1]
[3]	[3]	[4]	[5]	[0]	[1]	[2]
[4]	[4]	[5]	[0]	[1]	[2]	[3]
[5]	[5]	[0]	[1]	[2]	[3]	[4]

·	[0]	[1]	[2]	[3]	[4]	[5]
[0]	[0]	[0]	[0]	[0]	[0]	[0]
[1]	[0]	[1]	[2]	[3]	[4]	[5]
[2]	[0]	[2]	[4]	[0]	[2]	[4]
[3]	[0]	[3]	[0]	[3]	[0]	[3]
[4]	[0]	[4]	[2]	[0]	[4]	[2]
[5]	[0]	[5]	[4]	[3]	[2]	[1]

Having looked at these examples, we proceed to the formal definition of a ring.

DEFINITION

A **ring** consists of a set R together with two operations, $+$ and $\cdot$, defined on R, that satisfy the following axioms:
1. $(R, +)$ is an abelian group.
2. If x and y are in R, so is $x \cdot y$.   (closure)
3. $(x \cdot y) \cdot z = x \cdot (y \cdot z)$ for all $x, y, z \in R$.   (associative)
4. $x \cdot (y + z) = x \cdot y + x \cdot z$ for all $x, y, z \in R$.   (left distributive)
5. $(y + z) \cdot x = y \cdot x + z \cdot x$ for all $x, y, z \in R$.   (right distributive)

We will refer to a ring either by the system notation $(R, +, \cdot)$ or by the set R when the operations have been specified separately.

Each of our examples in this chapter is a ring. To study the properties that are common to all rings, we will investigate the consequences of our defining axioms. We will prove theorems which follow from the ring axioms and therefore apply to all rings.

EXERCISES
1. Show that the even integers are closed under multiplication and satisfy both distributive laws.
2. Show that for X, Y, and Z in $M_{2 \times 2}$

   $(Y + Z)X = YX + ZX$.
3. If Z is the set of all integers, show that $(Z, +, \cdot)$ is a ring.
4. Show that the rational numbers form a ring under addition and multiplication.
5. Perform the following multiplications:
   a. $[6] \cdot [7]$ in Z_{13}
   b. $[6] \cdot [7]$ in Z_{14}
   c. $[6] \cdot [7]$ in Z_{15}
   d. $[6] \cdot [7]$ in Z_{16}.
6. For $[a]$, $[b]$, and $[c]$ in Z_n, prove that

   $([b] + [c]) \cdot [a] = [b] \cdot [a] + [c] \cdot [a]$.

7. Let $(R, +)$ be an abelian group with identity 0. Define the operation $\cdot$ on R as follows: for $a, b \in R$, $a \cdot b = 0$. Show that $(R, +, \cdot)$ is a ring.
8. Show that the generalized associative law holds for multiplication in a ring. (See the discussion for groups.)
9. List three additional examples of rings.

Elementary ring theory

In this chapter we study some of the elementary properties of rings. Our approach will parallel the presentation of the theory of groups. We start with results that follow immediately from the axioms. Then we will study subrings and homomorphisms and their related topics.

BASIC CONCEPTS

A ring consists of a set R with two operations, $+$ and $\cdot$. Since R is an abelian group under the operation $+$, all the theorems proved in Part I hold for the additive structure of R. We repeat some of the basic results in the additive notation.

Notation: We will be using additive notation for the group $(R, +)$. The inverse of an element $x \in R$, under the operation $+$, will be written as $(-x)$ and referred to as the additive inverse of x. 0 will denote the identity of the group $(R, +)$ and will be called the additive identity of the ring $(R, +, \cdot)$. The expression $\underbrace{x + x + \cdots + x}_{n}$ will be denoted as nx; this "sum" in $(R, +)$ is analogous to the "product" $g \circ g \circ \cdots \circ g = g^n$ of an element g in a group $(G, \circ)$. Considering the operation of multiplication in R, we will denote $\underbrace{x \cdot x \cdot \cdots \cdot x}_{n}$ by x^n.

The next theorem consolidates facts that appeared in Part I either as theorems or exercises.

THEOREM 11.1
If $(R, +, \cdot)$ is a ring and $a, b, x, x_1, x_2, \ldots, x_n$ are any elements of R, then
1. a. $a + x = b + x \Rightarrow a = b$
   b. $x + a = x + b \Rightarrow a = b$ } (cancellation laws).
2. a. The additive identity 0 is unique.
   b. The additive inverse of any $x \in R$ is unique.
3. $x + x = x \Rightarrow x = 0$.
4. a. The sum $x_1 + x_2 + \cdots + x_n$ is uniquely defined (generalized associative law for addition).
   b. The product $x_1 \cdot x_2 \cdot \cdots \cdot x_n$ is uniquely defined (generalized associative law for multiplication).
5. For any integer n, $-(nx) = n(-x)$.
6. For any integers n and m
   a. $nx + mx = (n + m)x$
   b. $n(mx) = (nm)x$.
7. $-(-x) = x$.
8. $n(a + b) = na + nb$ for any integer n.

The proof of each part of this theorem can be found in Chapter 4 either as the proof to a theorem or corollary or as an exercise. For example, Theorem 11.1, part 6, is analogous to Theorem 4.7.

We will now prove some additional results that give us information as to how the additive and multiplicative structures of a ring interact.

THEOREM 11.2
Let $(R, +, \cdot)$ be a ring with additive identity 0. Then for any element $x \in R$,
$x \cdot 0 = 0 \cdot x = 0$.

Proof: Since 0 is the additive identity of R,

$0 + 0 = 0$.

Choose any element $x \in R$. We see that

$x \cdot (0 + 0) = x \cdot 0$.

Therefore

$x \cdot 0 + x \cdot 0 = x \cdot 0$

by the distributive law. Since $x \cdot 0 + 0 = x \cdot 0$ also, the cancellation laws guarantee that $x \cdot 0 = 0$.

The proof that $0 \cdot x = 0$ is done similarly. ∎

THEOREM 11.3
For any x, y in a ring $(R, +, \cdot)$,
1. $x \cdot (-y) = (-x)(y) = -(x \cdot y)$
2. $(-x) \cdot (-y) = x \cdot y$.

Proof of 1: Since $y + (-y) = 0$, we can write

$$x(y + (-y)) = x \cdot 0 = 0$$

or

$$xy + x(-y) = 0$$

by the distributive law. Since $(x \cdot y) + [-(x \cdot y)] = 0$ also, we can conclude that $-(x \cdot y) = x \cdot (-y)$ by the cancellation laws.

Similarly,

$$0 = (x + (-x)) \cdot y$$
$$= x \cdot y + (-x) \cdot y$$

so that $(-x) \cdot y = -(x \cdot y)$ also.

Proof of 2:

$$\begin{aligned}(-x) \cdot (-y) &= -[(x) \cdot (-y)] &&\text{(by part 1)} \\ &= -(-(x \cdot y)) &&\text{(by part 1 also)} \\ &= x \cdot y &&\text{(by Theorem 11.1, part 7)}\end{aligned}$$

completing our proof. ∎

EXERCISES

1. If $(R, +, \cdot)$ is a ring and $x \in R$, prove that $0 \cdot x = 0$.
2. Find the theorem, corollary, or exercise in Part I, Chapter 4, that is analogous to each part of Theorem 11.1.
3. If a ring $(R, +, \cdot)$ has an element e with the property $e \cdot x = x \cdot e = x$ for all $x \in R$, then e is called a multiplicative identity and R is called a ring with identity.
   a. Give an example of a ring with identity.
   b. Give an example of a ring without identity.
4. If $(R, +, \cdot)$ is a ring with a multiplicative identity e, prove that
   a. $(-e) \cdot x = -x$ for any $x \in R$.
   b. $(-e)(-e) = e$.
5. If $(R, +, \cdot)$ is a ring with the property that $x \cdot y = y \cdot x$ for any $x, y \in R$, then R is called a commutative ring.
   a. Give an example of a commutative ring.
   b. Give an example of a ring that is not commutative.
6. a. If R is a commutative ring, show that $(a \cdot b)^n = a^n \cdot b^n$ for any positive integer n.

b. Show that part a is not necessarily true in a noncommutative ring.
7. If R is a ring with identity e, show that $x \cdot x = e$ for every $x \in R$ implies that R is commutative.

SUBGROUPS AND IDEALS

Next, by analogy with what was useful for groups, we introduce the notion of a subring of a ring.

DEFINITION

Let $(R, +, \cdot)$ be a ring and let $S \subseteq R$. Then $(S, +, \cdot)$ is a **subring** of $(R, +, \cdot)$ if $(S, +, \cdot)$ is a ring.

Note that this definition means that (1) $(S, +)$ is a group and therefore, a subgroup of $(R, +)$; (2) $(S, \cdot)$ is closed under $\cdot$; and (3) the other ring axioms hold for $(S, +, \cdot)$. The next theorem states that the other axioms will automatically hold.

THEOREM 11.4

Let $(R, +, \cdot)$ be a ring and $S \subseteq R$. If $(S, +)$ is a subgroup of $(R, +)$ and S is closed under $\cdot$, then $(S, +, \cdot)$ is a subring of $(R, +, \cdot)$.

Proof: We have to check that $(S, +, \cdot)$ satisfies all the ring axioms.

a. Since $(S, +)$ is a subgroup of $(R, +)$, $(S, +)$ is a group. Since every subgroup of an abelian group is abelian, $(S, +)$ is an abelian group.

b. $(S, \cdot)$ is closed by the hypothesis of the theorem.

c. To show that $(S, \cdot)$ is associative, let x, y, and z be any elements of S. Then $x, y, z \in R$ and $(x \cdot y) \cdot z = x \cdot (y \cdot z)$ by the associative axiom in the definition of a ring.

d. To show that the distributive laws hold, consider any $x, y, z \in S$. Then $x, y, z \in R$ and by the distributive properties of R, both

$$x \cdot (y + z) = x \cdot y + x \cdot z$$

and

$$(y + z) \cdot x = y \cdot x + z \cdot x.$$

Since S satisfies all the axioms in the definition of a ring, $(S, +, \cdot)$ is a ring and therefore a subring of $(R, +, \cdot)$. ∎

COROLLARY

If R is a ring and S is a nonempty subset of R, and if for every $x, y \in S$,
1. $x + (-y) = x - y \in S$ and
2. $x \cdot y \in S$

then S is a subring of R.

Proof: If we translate Theorem 5.2 of Part I into additive notation, we see that a nonempty subset $S \subseteq R$ is a subgroup of $(R, +)$ if for every $x, y \in S$,

$$x + (-y) = (x - y) \in S.$$

Therefore the hypotheses of Theorem 11.4 are satisfied and S is a subring of R if $(x - y) \in S$ and $x \cdot y \in S$. ∎

EXAMPLE 11.1

In the ring of integers $(Z, +, \cdot)$ we will use the corollary to Theorem 11.4 to show that the even integers E form a subring of Z. $0 \in E$ so $E \neq \emptyset$. Choose $2m, 2n \in E$.

a. Then $2m + (-2n) = 2(m - n)$ which is in E. Therefore $(E, +)$ is a subgroup of $(Z, +)$.

b. Also $(2m)(2n)$ is also in E since it is 2 times the integer $2mn$. Therefore E is closed under $\cdot$.

By Theorem 11.4 and its corollary, $(E, +, \cdot)$ is a subring of $(Z, +, \cdot)$.

Note that some properties of a ring are not preserved by its subrings.

EXAMPLES 11.2

Consider the ring $(Z_6, +, \cdot)$ where $Z_6 = \{0, 1, 2, 3, 4, 5\}$ and the operations are addition and multiplication modulo 6.

The element 1 acts as a multiplicative identity for this ring.

1. The subset $\{0, 2, 4\}$ forms a subring with addition and multiplication tables

+	0	2	4		·	0	2	4
0	0	2	4		0	0	0	0
2	2	4	0		2	0	4	2
4	4	0	2		4	0	2	4

By inspection, we can see that the subring $\{0, 2, 4\}$ has no multiplicative identity.

2. The subset $\{0, 3\} \subseteq Z_6$ is a subring with addition and multiplication tables

+	0	3		·	0	3
0	0	3		0	0	0
3	3	0		3	0	3

By inspection we can see that the integer 3 acts as a multiplicative identity in the subring. That is, the subring $\{0, 3\}$ has a multiplicative identity that is different from the identity in the ring Z_6.

Elementary ring theory 145

We would like to point out that many books require that all subrings of a ring with identity e must also contain e. We do not need that requirement for the material covered in this text and therefore do not make that requirement. However, the student is warned to read the definition of a subring in other texts with an eye for this difference.

We note that for any subring S of a ring R, the group $(S, +)$ is a subgroup of $(R, +)$. Since $(R, +)$ is an abelian group, S is a normal subgroup of R. We can consider $R|S$, the set of all cosets of R in S. That is,

$$R|S = \{S + a \mid a \in R\}.$$

(Recall that the group operation in R is $+$, so that all cosets are of the form $S + a$ for some element $a \in R$.) Since R and S are groups under the operation $+$ and $S \triangleleft R$, Theorem 6.9 of Part I and the following discussion proves that the addition of cosets defined by

$$(S + a) + (S + b) = S + (a + b)$$

is well defined. Theorem 6.10 states that $R|S$ is a group under this operation.

However, $R|S$ is not necessarily a ring since there is no multiplicative operation defined on it. The operation defined by

$$(S + a) \cdot (S + b) = S + ab$$

is not necessarily well defined on any subring. We need an extra condition.

DEFINITION
If R is a ring, then a subring $I \subseteq R$ is called an **ideal** if for every $r \in R$ and $i \in I$,

$$i \cdot r \in I \quad \text{and} \quad r \cdot i \in I.$$

EXAMPLES 11.3
1. In the ring of integers Z, the even integers E form a subring. We will show that E is an ideal of Z. Choose any element $n \in Z$ and $2m \in E$. Consider the products

$$n(2m) = 2nm$$

and

$$(2m)n = 2mn.$$

Since both products are even, they both belong to E and E is an ideal of Z.

2. The reals $\mathbb{R}$ form a ring and the integers Z form a subring of $\mathbb{R}$. However, if we take the element $\sqrt{2}$ in $\mathbb{R}$ and the element 3 in Z, the product

$3\sqrt{2} \notin Z.$

Therefore Z is not an ideal of the reals.

3. In the ring Z_6, the subset $\{0, 3\}$ is a subring. We check the products of elements in $\{0, 3\}$ with elements of Z_6 under multiplication modulo 6.

$0 \cdot 0 \equiv 0 \pmod{6}$	$3 \cdot 0 = 0 \cdot 3 \equiv 0 \pmod{6}$
$0 \cdot 1 = 1 \cdot 0 \equiv 0 \pmod{6}$	$3 \cdot 1 = 1 \cdot 3 \equiv 3 \pmod{6}$
$0 \cdot 2 = 2 \cdot 0 \equiv 0 \pmod{6}$	$3 \cdot 2 = 2 \cdot 3 \equiv 0 \pmod{6}$
$0 \cdot 3 = 3 \cdot 0 \equiv 0 \pmod{6}$	$3 \cdot 3 \equiv 3 \pmod{6}$
$0 \cdot 4 = 4 \cdot 0 \equiv 0 \pmod{6}$	$3 \cdot 4 = 4 \cdot 3 \equiv 0 \pmod{6}$
$0 \cdot 5 = 5 \cdot 0 \equiv 0 \pmod{6}$	$3 \cdot 5 = 5 \cdot 3 \equiv 3 \pmod{6}$

Since all the products are either 0 or 3, $\{0, 3\}$ is an ideal of Z_6.

4. In any ring R, both $\{0\}$ and R itself are ideals. These results are left as exercises.

We will show that if I is an ideal of R, then the set $R|I$ forms a ring under the appropriate operations. In this way, the concept of ideals in rings is analogous to the concept of normal subgroups in groups. In the next section we will see that ideals of rings and normal subgroups share another important property.

THEOREM 11.5

If R is a ring and I is an ideal of R, then the operation $\cdot$, defined on $R|I$ by

$(I + a) \cdot (I + b) = I + a \cdot b$

is well defined.

Proof: To show that $\cdot$ is well defined, we need to show that the definition is independent of the particular coset representatives chosen. That is, if $x \in I + a$ and $y \in I + b$, then for the definition to be consistent, it must be true that

$(I + x) \cdot (I + y) = (I + a)(I + b)$

or

$I + x \cdot y = I + a \cdot b.$

Elementary ring theory

Equivalently, we have to verify that if $x \in I + a$ and $y \in I + b$, then $x \cdot y \in I + a \cdot b$.

If $x \in I + a$ and $y \in I + b$, then $x = i + a$ and $y = j + b$ for some elements $i, j \in I$. Therefore

$$x \cdot y = (i + a) \cdot (j + b)$$
$$= i \cdot j + i \cdot b + a \cdot j + a \cdot b.$$

Since I is an ideal, the elements $i \cdot j$, $i \cdot b$, and $a \cdot j$ belong to I and their sum is also in I. Let

$$k = i \cdot j + i \cdot b + a \cdot j \in I$$

and then

$$x \cdot y = k + a \cdot b$$

or

$$x \cdot y \in I + a \cdot b. \blacksquare$$

Note that in this proof, it is necessary that I is an ideal of R and not just a subring. For every a, b in R and i, j in I, we needed the fact that $i \cdot b$ and $a \cdot j$ are elements of I.

THEOREM 11.6
Let R be a ring and let I be an ideal of R. Let $R|I = \{I + a \mid a \in R\}$. Then $R|I$ is a ring under the operations $+$ and $\cdot$ defined by

$$(I + a) + (I + b) = I + (a + b)$$

and

$$(I + a) \cdot (I + b) = I + a \cdot b.$$

Proof: We will verify each of the ring axioms.

1. Since $(I, +)$ is a normal subgroup of $(R, +)$, $(R|I, +)$ is a group. Note that

$$(I + a) + (I + b) = I + (a + b)$$
$$= I + (b + a)$$
$$= (I + b) + (I + a)$$

and $R|I$ is commutative under $+$.

2. By definition, $(I + a) \cdot (I + b)$ is an element of $R|I$ so that $R|I$ is closed under the operation $\cdot$.

3. To show that $R|I$ is associative under $\cdot$, we choose $I+a$, $I+b$, $I+c$ in $R|I$. Then

$$\begin{aligned}(I+a)\cdot[(I+b)\cdot(I+c)] &= (I+a)(I+b\cdot c)\\ &= I+a\cdot(b\cdot c)\\ &= I+(a\cdot b)\cdot c \quad \text{(since } R \text{ is associative)}\\ &= (I+a\cdot b)\cdot(I+c)\\ &= [(I+a)\cdot(I+b)]\cdot(I+c).\end{aligned}$$

4. To show that the distributive laws are satisfied by this system, we again consider elements $I+a$, $I+b$, and $I+c$ in $R|I$.

$$\begin{aligned}(I+a)\cdot[(I+b)+(I+c)] &= (I+a)\cdot[I+(b+c)]\\ &= I+a\cdot(b+c)\\ &= I+(a\cdot b+a\cdot c) \quad \text{(why?)}\\ &= (I+a\cdot b)+(I+a\cdot c)\end{aligned}$$

which is the first distributive law in $R|I$. The fact that

$$[(I+b)+(I+c)]\cdot(I+a) = (I+b\cdot a)+(I+c\cdot a)$$

is proved similarly. ∎

This theorem is an exact analog to the corresponding one for groups and asserts that ideals play the same role in ring theory as normal subgroups do in group theory.

DEFINITION
The ring $R|I$ is called the **quotient ring** of R by I.

We conclude this section by presenting some related definitions and examples.

DEFINITIONS
If R is a ring:
1. A **left ideal** of R is a subring A with the property that for every $r \in R$ and $a \in A$, $r \cdot a \in A$.
2. A **right ideal** of R is a subring B such that for every $b \in B$ and $r \in R$, $b \cdot r \in B$.

EXAMPLE 11.4
Consider the ring $M_{2\times 2}$ of all 2×2 matrices under matrix addition and multiplication. The set $A = \left\{ \begin{pmatrix} 0 & x \\ 0 & y \end{pmatrix} \middle| x, y \in \mathbb{R} \right\}$ is a left ideal of $M_{2\times 2}$

Elementary ring theory

since for any $\begin{pmatrix} a & b \\ c & d \end{pmatrix} \in M_{2 \times 2}$, the product

$$\begin{pmatrix} a & b \\ c & d \end{pmatrix}\begin{pmatrix} 0 & x \\ 0 & y \end{pmatrix} = \begin{pmatrix} 0 & ax+by \\ 0 & cx+dy \end{pmatrix} \in A.$$

Also, the set $B = \left\{ \begin{pmatrix} x & y \\ 0 & 0 \end{pmatrix} \middle| x, y \in \mathbb{R} \right\}$ is a right ideal.

For any commutative ring R ($a \cdot b = b \cdot a$ for every $a, b \in R$), every left ideal is a right ideal.

DEFINITIONS

1. $\{0\}$ and R itself are called the **trivial ideals** of a ring R.
2. A ring without any nontrivial ideals is called a **simple** ring.

EXAMPLE 11.5
$M_{2 \times 2}$ is a simple ring.

EXERCISES

1. Let Z be the ring of integers and $H_n = \{kn \mid k \in Z\}$. Show that H_n is a subring of Z.
2. Show that $\left\{ \begin{pmatrix} a & 0 \\ 0 & 0 \end{pmatrix} \middle| a \in \mathbb{R} \right\}$ is a subring of $M_{2 \times 2}$.
3. Give two additional examples of a ring with identity and a subring without identity.
4. a. Give an example of a noncommutative ring with a commutative subring.
   b. Is it possible to have a commutative ring with a noncommutative subring?
5. If R is a ring, show that R and $\{0\}$ are ideals of R.
6. Show that the ring of real numbers is simple.
7. If R is a ring with a multiplicative identity 1, show that any ideal that contains 1 must be R itself.
8. a. Show that $\left\{ \begin{pmatrix} x & y \\ 0 & 0 \end{pmatrix} \middle| x, y \in \mathbb{R} \right\}$ is a right ideal of $M_{2 \times 2}$.
   b. Give examples of another left ideal and another right ideal of $M_{2 \times 2}$.
   c. Show that $M_{2 \times 2}$ is simple.
9. Let $(R, +, \cdot)$ be a ring such that $(R - \{0\}, \cdot)$ is a group. Show that R is simple.
10. Let $(R, +, \cdot)$ be a ring with the property that $a \cdot b = 0$ for every $a, b \in R$. Show that any subgroup of $(R, +)$ is an ideal.
11. a. Show that the intersection of ideals is an ideal.
    b. Prove the analogous results for right and left ideals.
12. Consider the set $K_{2 \times 2}$ of all 2×2 matrices with integer entries.
    a. Show that $K_{2 \times 2}$ is a subring of $M_{2 \times 2}$.
    b. Show that $K_{2 \times 2}$ is not simple.

13. If R is a ring and A and B are subsets of R, define
$$AB = \{a_1b_1 + a_2b_2 + \cdots + a_nb_n | a_i \in A, b_j \in B, i, j = 1, \ldots, n, n \in Z^+\}.$$
   a. If $A = \{-1, 2, \frac{3}{2}\}$ $B = \{2, 3\}$, $A, B \subseteq \mathbb{R}$, what is AB?
   b. Show that $I \subseteq R$ is an ideal if and only if $IR \subseteq I$ and $RI \subseteq I$.
   c. Prove analogous results for right and left ideals.
14. Let $\mathbb{R}$ be the ring of real numbers, let Z be the subring of integers, and let $\mathbb{R}|Z$ be the set of cosets of Z in $\mathbb{R}$. Give an example to show that the multiplication $(Z + a) \cdot (Z + b) = Z + ab$ is not well defined.
15. Let Z be the ring of integers and let H_n be the subring of integers divisible by n. What is $Z|H_n$? Write out addition and multiplication tables for $n = 3, 5, 6$.
16. Let $K_{2 \times 2}$ be the subring of $M_{2 \times 2}$ that is described in Exercise 12. Let $\mathcal{H}_n$ be the subset of matrices whose entries are divisible by n.
   a. Show $\mathcal{H}_n$ is an ideal in $K_{2 \times 2}$.
   b. Write out addition and multiplication tables for $K_{2 \times 2}|\mathcal{H}_n$.
17. Let I be an ideal of a ring R. Prove there is a bijection (1–1 correspondence) between ideals of $R|I$ and ideals of R which contain I.

HOMOMORPHISMS

Another concept which was very important in our study of groups was that of homomorphisms and the related concepts of isomorphisms and automorphisms. In studying groups we defined a homomorphism from a group G into a group H as a mapping or function from G to H which preserves the group operations. Rings have two operations, but the basic idea is the same.

DEFINITION
If R and S are rings, a **ring homomorphism** from R to S is a function $\alpha : R \to S$ with the property that for all $x, y \in R$,

1. $\alpha(x + y) = \alpha(x) + \alpha(y)$ and
2. $\alpha(x \cdot y) = \alpha(x) \cdot \alpha(y)$.

EXAMPLE 11.6
Let Z be the ring of integers, and for any integer $n > 1$, let Z_n be the ring of integers modulo n. These are the rings defined in Example 10.5 except that we denote an element of Zn as i instead of $[i]$. For any integer k in Z, there is an element $a \in Z_n$ such that $k \equiv a(\bmod n)$. We define a function $\alpha : Z \to Z_n$ by

$\alpha(k) = a$.

Elementary ring theory

To show that α is a homomorphism from Z to Z_n, we have to show for any two integers i and j,

1. $\alpha(i + j) = \alpha(i) + \alpha(j)$
2. $\alpha(i \cdot j) = \alpha(i) \cdot \alpha(j)$.

1. Suppose $i \equiv a(\bmod n)$ and $j \equiv b(\bmod n)$. Then $n|(i - a)$ and $n|(j - b)$. Therefore, since $(i - a) + (j - b) = (i + j) - (a + b)$, $n|[(i + j) - (a + b)]$ and $i + j \equiv (a + b) \bmod (n)$. We see that

$$\alpha(i + j) = a + b$$
$$= \alpha(i) + \alpha(j)$$

completing the proof of 1.

2. Again, suppose $i \equiv a(\bmod n)$ and $j \equiv b(\bmod n)$. Then $n|(i - a)$ and $n|(j - b)$ or $i - a = nk$ and $j - b = nl$ for some integers k and l. Solving, we get

$$i = a + nk \quad \text{and} \quad j = b + nl.$$

The product

$$ij = (a + nk)(b + nl)$$
$$= ab + nkb + nla + n^2kl$$
$$= ab + n(kb + la + nkl)$$

and we see that $n|(ij - ab)$ or $ij \equiv ab(\bmod n)$. Therefore

$$\alpha(ij) = ab$$
$$= \alpha(i) \cdot \alpha(j)$$

which completes the proof that α is a ring homomorphism. Note that we have defined a different α for each integer n. That is, for $n = 2$, we have a homomorphism from Z to Z_2; for $n = 3$, there is a homomorphism from Z to Z_3, and so on. All of these homomorphisms have been defined by considering the general case.

Note: A ring homomorphism $\alpha: R \to S$ is a group homomorphism on the additive structure of the ring R. Therefore the following results are automatically true by Theorem 7.6:
1. $\alpha(0) = 0$
2. $\alpha(-x) = -\alpha(x)$ for every x in R.

Before we continue our study of ring homomorphisms, we will define and make some observations about ring isomorphisms and automorphisms.

DEFINITION
Let R and S be rings.
1. A **ring isomorphism,** $\alpha: R \to S$, is a bijective (1–1 and onto) ring homomorphism.
2. If $\alpha: R \to S$ is an isomorphism, the rings R and S are said to be **isomorphic.**
3. An **automorphism** of a ring R is an isomorphism that maps R onto itself.

EXAMPLES 11.7
1. Let $Z_2 = \{0, 1\}$ be the ring of integers modulo 2 and let $S = \{x, y\}$ be the ring with addition and multiplication tables

+	x	y		$\cdot$	x	y
x	x	y		x	x	x
y	y	x		y	x	y

The mapping $\alpha: Z_2 \to S$ defined by

$\alpha(0) = x$

$\alpha(1) = y$

is an isomorphism and Z_2 and S are isomorphic rings. As with groups, isomorphic rings can be interpreted as the same ring except for the notation used.

2. Consider the ring $Z_3 = \{0, 1, 2\}$ of integers modulo 3. The function $\alpha: Z_3 \to Z_3$ defined by

$\alpha(0) = 0$

$\alpha(1) = 2$

$\alpha(2) = 1$

is an automorphism of Z_3.

3. If R is any ring, the identity map $i: R \to R$, defined by $i(r) = r$ for every r in R, is an automorphism of R. This is identical to a result we obtained with group theory.

THEOREM 11.7
The set of all automorphisms of a ring forms a group under the operation of composition.

The proof is left as an exercise. It is essentially the same as the proof of Theorem 7.4 in Part I. Note that there is no ring structure defined on the set of all automorphisms of a ring.

Elementary ring theory

We seem to be developing the pattern of trying to study concepts for rings that are similar to the ones we studied for groups. When we studied group homomorphisms we found that the image of the homomorphism was a subgroup of the codomain and the kernel of the homomorphism was a normal subgroup of the domain. Now we ask what happens if we have a ring homomorphism $\alpha: R \to S$. We make some definitions first.

DEFINITIONS
Let R and S be rings and let $\alpha: R \to S$ be a ring homomorphism.
1. The **image** of $\alpha = \text{Im}(\alpha) = \{\alpha(r) \mid r \in R\}$.
2. The **kernel** of $\alpha = \ker(\alpha)$
$$= \{r \mid \alpha(r) = 0, \text{ the additive identity in } S\}.$$

EXAMPLE 11.8
In the ring homomorphism $\alpha: Z \to Z_n$, defined by $\alpha(k) = a \in Z_n$ where $k \equiv a \pmod{n}$,

$\text{Im}(\alpha) = Z_n$.

Since for every $j \in Z_n$, $0 \leq j \leq n - 1$, the integer $j \in Z$ has the property that $\alpha(j) = j$ since $j \equiv j \pmod{n}$.

Also, we see that

$\ker(\alpha) = \{kn \mid k \in Z\}$

since for every integer k, $kn \equiv 0 \pmod{n}$ and $\alpha(kn) = 0$.

THEOREM 11.8
If $\alpha: R \to S$ is a ring homomorphism,
1. $\text{Im}(\alpha)$ is a subring of S, and
2. $\ker(\alpha)$ is a subring of R.

Proof: We will use the corollary to Theorem 11.4 to prove both parts of the theorem.

Proof of 1: First we note that $\text{Im}(\alpha)$ is not empty since for $0 \in R$, $\alpha(0) = 0 \in \text{Im}(\alpha)$.

Let x and y be any elements of $\text{Im}(\alpha)$. There are elements $a, b \in R$ such that

$\alpha(a) = x$ and $\alpha(b) = y$.

We will show that $x - y$ and $x \cdot y$ are elements of $\text{Im}(\alpha)$ by showing that they have preimages in R. Since α is a homomorphism,

$x - y = \alpha(a) - \alpha(b)$
$= \alpha(a) + \alpha(-b)$
$= \alpha(a - b) \in \text{Im}(\alpha)$.

Also,
$$x \cdot y = \alpha(a) \cdot \alpha(b)$$
$$= \alpha(a \cdot b) \in \text{Im}(\alpha)$$
and Im(α) is a subring of S.

Proof of 2: ker(α) $\neq \emptyset$ since the additive identity of R maps to the 0 in S.

Choose elements x and y in ker(α) ($\alpha(x) = 0$ and $\alpha(y) = 0$). We will consider the images of $x - y$ and $x \cdot y$.

$$\alpha(x - y) = \alpha(x + (-y))$$
$$= \alpha(x) + \alpha(-y)$$
$$= \alpha(x) - \alpha(y)$$
$$= 0 - 0 = 0.$$
$$\alpha(x \cdot y) = \alpha(x) \cdot \alpha(y)$$
$$= 0 \cdot 0 = 0.$$

Since both $x - y$ and $x \cdot y$ map to the 0 in S, $(x - y) \in \text{ker}(\alpha)$ and $x \cdot y \in \text{ker}(\alpha)$ so that ker(α) is a subring of R. ∎

In studying group homomorphisms, we found that the kernel was not only a subgroup but a normal subgroup. Since an ideal of a ring is the ring-theoretic counterpart of a normal subgroup in the study of quotient groups and rings, we might expect that the analogy would carry through here.

THEOREM 11.9
Let R and S be rings and let $\alpha: R \to S$ be a ring homomorphism; then ker(α) is an ideal of R.

Proof: We need to show for any $i \in \text{ker}(\alpha)$ ($\alpha(i) = 0$) and $r \in R$, that $r \cdot i$ and $i \cdot r$ are elements of ker(α). We consider the images of $r \cdot i$ and $i \cdot r$.

$$\alpha(r \cdot i) = \alpha(r) \cdot \alpha(i) \quad \text{(since } \alpha \text{ is a homomorphism)}$$
$$= \alpha(r) \cdot 0 \quad \text{(since } i \in \text{ker}(\alpha))$$
$$= 0 \quad \text{(by Theorem 2.2)}$$

and
$$\alpha(i \cdot r) = \alpha(i) \cdot \alpha(r)$$
$$= 0 \cdot \alpha(r)$$
$$= 0.$$

(The reasons for the second string of equalities are the same as for the first string.) Since for every $r \in R$, $i \in \ker(\alpha)$, both $i \cdot r$ and $r \cdot i$ map to 0, $i \cdot r \in \ker(\alpha)$ and $r \cdot i \in \ker(\alpha)$ and $\ker(\alpha)$ is an ideal. ∎

Note that the theorem tells us that the integers can never be the kernel of a homomorphism from the real numbers to another ring. (Why?)
We can now state and prove an analog of Theorems 7.10 and 7.11.

THEOREM 11.10
1. Let α be a surjective homomorphism from a ring R onto a ring S. Then $R|\ker(\alpha)$ is isomorphic with S.
2. If I is any ideal of a ring R, then the mapping $\alpha: R \to R|I$, defined by $\alpha(x) = I + x$, is a surjective homomorphism with kernel I.

Proof of 1: The proof depends heavily on the proof of Theorem 7.10. Consider the mapping β from $R|\ker(\alpha)$ to S defined by $\beta(I + x) = \alpha(x)$. The proof of Theorem 7.10 tells us that β is well defined, injective, surjective, and a group homomorphism under the additive operations of R and S. Therefore we only need to verify that β is also a ring homomorphism. That is, we will check to see that the multiplicative operation is preserved. Choose $\ker(\alpha) + x$ and $\ker(\alpha) + y$ in $R|\ker(\alpha)$; then

$$\beta((\ker(\alpha) + x)(\ker(\alpha) + y)) = \beta(\ker(\alpha) + x \cdot y)$$
$$= \alpha(x \cdot y)$$
$$= \alpha(x) \cdot \alpha(y)$$
$$= \beta(\ker(\alpha) + x) \cdot \beta(\ker(\alpha) + y)$$

which completes the proof. ∎

We leave the proof of part 2 as an exercise.

EXERCISES
1. Let α be a homomorphism from a ring R to a ring S.
   a. Show that α maps the additive identity of R onto the additive identity of S.
   b. Show that for any $x \in R$, $\alpha(-x) = -\alpha(x)$.
2. If $(R, +, \cdot)$ is a ring, define a new multiplication $\circ$ on R by $x \circ y = 0$ for every $x, y \in R$. We will call the system $(R, +, \circ)$ the **trivial ring** and denote it by R°. Show that the mapping $\alpha: R \to R^\circ$ defined by $\alpha(x) = x$ is not a homomorphism unless $(R, +, \cdot)$ is the trivial ring itself.
3. Verify that the kernel of the map $\alpha: Z \to Z_n$ is an ideal of Z.
4. Is it possible to define a homomorphism with domain $\mathbb{R}$ and kernel to be the set of rational numbers? Why or why not?

5. Prove that the automorphisms of a ring form a group.
6. Prove that isomorphic rings have isomorphic automorphism groups.
7. Let Q be the ring of rational numbers. Show that any automorphism α of Q which leaves every integer fixed ($\alpha(n) = n$) must be the identity automorphism.
8. Find a ring R, a subring S, and an automorphism $\alpha: R \to R$ such that $\alpha(s) = s$ for every $s \in S$ but that α is not the identity automorphism.
9. Prove part 2 of Theorem 11.10.

CHARACTERISTIC OF A RING

As a last concept in this chapter, before we consider special classes of rings, we will look at the notion of the characteristic of a ring.

Let $(R, +, \cdot)$ be a ring and let x be any element of R. We want to investigate repeated sums of x's. Recall that we denote the repeated sum $\underbrace{x + x + \cdots + x}_{n}$ by nx where n is a positive integer. This notation leads to the following result.

LEMMA

If R is a ring and $x, y \in R$, then for any positive integer n,
$n(xy) = (nx)y = x(ny)$.

$\textit{Proof}: n(xy) = \underbrace{xy + xy + \cdots + xy}_{n}$

$\qquad\qquad = \underbrace{(x + x + \cdots + x)}_{n} y \qquad \text{(why?)}$

$\qquad\qquad = (nx)y.$

Similarly,

$n(xy) = \underbrace{xy + xy + \cdots + xy}_{n}$

$\qquad = x\underbrace{(y + y + \cdots + y)}_{n}$

$\qquad = x(ny).$ ∎

Let us look at the repeated sums of elements of Z_3 as an example.

EXAMPLE 11.9
The elements of Z_3 are 0, 1, and 2, and the tables for $+$ and $\cdot$ are

+	0	1	2
0	0	1	2
1	1	2	0
2	2	0	1

$\cdot$	0	1	2
0	0	0	0
1	0	1	2
2	0	2	1

The repeated sums are

$1 \cdot 0 = 0 = 0$

$2 \cdot 0 = 0 + 0 = 0$

$3 \cdot 0 = 0 + 0 + 0 = 0$

$1 \cdot 1 = 1 = 1$

$2 \cdot 1 = 1 + 1 = 2$

$3 \cdot 1 = 1 + 1 + 1 = 0$

$1 \cdot 2 = 2 = 2$

$2 \cdot 2 = 2 + 2 = 1$

$3 \cdot 2 = 2 + 2 + 2 = 0$

We note that for every $x \in Z_3$, $3x = 0$.

DEFINITIONS
Let R be a ring.

1. If there is a least integer $k > 0$, such that $kx = 0$ for all $x \in R$, then k is called the **characteristic** of R.

2. If there is no integer k such that $kx = 0$ for every $x \in R$, then R is said to have **characteristic 0**.

EXAMPLES 11.10
1. As we saw above, the characteristic of Z_3 is 3. In fact, this result can be generalized to the fact that for every positive integer m, the characteristic of Z_m is m.

2. $\mathbb{R}$, the ring of real numbers, is a ring with characteristic 0.

We will present some methods of establishing the characteristics of some particular types of rings.

If a ring $(R, +, \cdot)$ has an element e such that $e \cdot x = x \cdot e = x$ for all x in R, then e is called the multiplicative identity for R and R is called a **ring with identity**. The next result states a method for finding the characteristic of a ring with identity.

THEOREM 11.11
Let R be a ring with identity e; then the characteristic of R is k if and only if k is the least positive integer such that $ke = 0$.

Proof: 1. Assume R has characteristic k. Then $ke = 0$. We need to show that k is the smallest integer for which this happens. By contradiction, assume $me = 0$ for some integer $m < k$. Then for any $x \in R$,

$$mx = m(e \cdot x)$$
$$= (me) \cdot x$$
$$= 0 \cdot x$$
$$= 0.$$

This contradicts the hypothesis that R has characteristic k. Therefore, if the characteristic of R is k, k is the smallest positive integer for which $ke = 0$.

2. Finally, if k is the smallest integer such that $ke = 0$, then for every $x \in R$,

$$kx = k(e \cdot x)$$
$$= (ke) \cdot x$$
$$= 0 \cdot x$$
$$= 0.$$

Therefore k is the characteristic of R. ∎

Note that the characteristic of a ring with identity e is the additive order of e.

As we shall see later, the notion of the characteristic of a ring is of prime importance in the study of rings.

We will conclude this chapter by introducing two special classes of rings and looking at a result dealing with their characteristics.

DEFINITIONS

1. An **integral domain** is a ring $(R, +, \cdot)$ with the following properties:
a. $a \cdot b = b \cdot a$ for every $a, b \in R$.
b. There is an $e \in R$ such that

$a \cdot e = e \cdot a = a$ for every $a \in R$.

c. If $a, b \in R$ and $a \cdot b = 0$, then either $a = 0$ or $b = 0$.
2. A **field** is an integral domain $(R, +, \cdot)$ with the additional property that the system $(R - \{0\}, \cdot)$ is an abelian group.

EXAMPLES 11.11

1. The ring of integers $(Z, +, \cdot)$ is an integral domain since
a. $(Z, +, \cdot)$ is a ring
b. multiplication of integers is commutative

c. 1 acts as a multiplicative identity and
d. $i \cdot j = 0$ implies $i = 0$ or $j = 0$.

The ring of integers is not a field since no integers except 1 and -1 have multiplicative inverses. Therefore $(Z - \{0\}, \cdot)$ is not a group.

2. The ring of real numbers $(\mathbb{R}, +, \cdot)$ is an integral domain and a field. This was shown in Examples 1.1 and 1.2.

3. The fact that the ring $(Z_5, +, \cdot)$ is a field was shown in Chapter 6. However, $(Z_6, +, \cdot)$ is not even an integral domain since $3 \cdot 2 = 0$ in Z_6. We will develop this concept further in the exercises.

In an integral domain, we have the following result on characteristics.

THEOREM 11.12
Let $(R, +, \cdot)$ be an integral domain. If the characteristic of R is k, then either $k = 0$ or k is a prime number.

Proof: Case 1: If $k = 0$, the theorem holds.

Case 2: If $k \neq 0$, we will prove the theorem by contradiction. Let k be a composite integer, $k = mn$ for some $m, n \in Z$. Since R is an integral domain, R has a multiplicative identity e. Then $ke = 0$ and by Theorem 11.11, k is the smallest positive integer for which this is true. But if $k = mn$, then

$0 = (mn)e$

$= (me)(ne)$ (why?)

and the fact that R is an integral domain implies that $me = 0$ or $ne = 0$. This contradicts Theorem 11.11. We can conclude that the characteristic of R is prime if it is not 0. ∎

COROLLARY
The characteristic of a field is either 0 or a prime number.

Proof: A field is an integral domain and Theorem 11.12 holds. ∎

This ends our study of elementary ring theory. We will be using the results of this chapter as we continue our study of rings.

EXERCISES

1. For any positive integer n, show that the characteristic of Z_n is n.
2. Show that Z_n is an integral domain if and only if n is prime.
3. Let R be a ring with identity and let S be any ring. Let $\alpha : R$ onto S be a surjective homomorphism.
   a. Show that S has an identity.
   b. Show that if the characteristic of R is n and the characteristic of S is m, then $m|n$.

4. Let $S = \left\{ \begin{pmatrix} a & b \\ c & d \end{pmatrix} \middle| a, b, c, d \in Z_n \right\}$. Show that the characteristic of S is n.
5. a. Show that Z_2, Z_3, and Z_5 are fields.
   b. Show that Z_4 and Z_6 are not fields.
   c. Prove that Z_n is a field if and only if n is prime.
6. Give an example of an integral domain that is not a field.
7. Prove that any finite integral domain is a field.
8. Do the cancellation laws hold for multiplication in an integral domain? Prove your result.
9. Prove that a field is an integral domain in which every nonzero element has a multiplicative inverse.
10. Show that the only ideals of a field are $\{0\}$ and the field itself. That is, show that every field is simple.
11. A ring D, for which $(D - \{0\}, \cdot)$ is a group, is called a **division ring**. (Note that D does not have to be commutative.) Which of the examples that we have studied in this part of the book are division rings?
12. Let $Q = \{(a, b, c, d) | a, b, c, d \in \mathbb{R}\}$. We define two operations on Q:

$$(a, b, c, d) + (e, f, g, h) = (a + e, b + f, c + g, d + h)$$

$$(a, b, c, d) \cdot (e, f, g, h) = (ae - bf - cg - dh, af + be + ch - dg,$$
$$ag - bh + ce + df, ah + bg - cf + de).$$

Show that $(Q, +, \cdot)$ is a division ring but not a field.

Special classes of rings

In this chapter we examine some important classes of rings. To a great extent, we will be studying rings with properties similar to well-known rings. A favorite technique in ring theory is to look at what seems to be a key property of a familiar ring and investigate all rings with that property.

In the last chapter we dealt with commutative rings, rings with identity, integral domains, and fields. In this chapter we will investigate Euclidean rings and polynomial rings.

EUCLIDEAN RINGS

One of the important features of the integers is the division algorithm. It was this result that allowed us to prove so many results in group theory.

In this chapter we will generalize this result to apply to a ring and study its consequences. First recall that the division algorithm states:

If n and m are nonzero integers, then there exist unique integers q and r such that $n = qm + r$, where $0 \leq r < |m|$.

This is a slight restatement which includes the corollary and results from the exercises.

DEFINITION
Let $(R, +, \cdot)$ be an integral domain. If there is mapping $\delta: R - \{0\} \to Z^+$ (from $R - \{0\}$ to the nonnegative integers) such that for any nonzero elements x, y in R,
1. $\delta(x) \leq \delta(xy)$ and
2. there exist elements q and r in R such that $x = q \cdot y + r$ where $r = 0$

or $\delta(r) < \delta(y)$, then the system $(R, +, \cdot)$ is called a **Euclidean ring** or a **Euclidean domain**.

EXAMPLES 12.1
1. For the integers Z, let $\delta(n) = |n|$, the absolute value of n. The definition then states that for any integers $x, y \neq 0$ that
   a. $|x| \leq |xy|$ and
   b. there exist integers q and r such that $x = q \cdot y + r$ where $0 \leq |r| < |y|$.

Note that the integers satisfy both conditions and the system $(Z, +, \cdot)$ is a Euclidean ring. Note that the division algorithm directly implies condition 2.

2. In the real numbers $\mathbb{R}$, let $\delta(x) = 1$ for every nonzero x in $\mathbb{R}$. Conditions 1 and 2 of the definition of a Euclidean ring translate to
   a. $1 \leq 1$ ($\delta(x) \leq \delta(xy)$ for every nonzero $x, y \in \mathbb{R}$)
   b. for $x, y \in \mathbb{R}$, $x, y \neq 0$, there are real numbers q and r such that $x = qy + r$ where either $r = 0$ or $1 < 1$ ($\delta(r) < \delta(y)$).

Since the statement $1 < 1$ is never true, $r = 0$ and this means that for every $x, y \in \mathbb{R}$, $x, y \neq 0$, there is a $q \in \mathbb{R}$ with

$$x = qy.$$

Because every nonzero element in $\mathbb{R}$ has a multiplicative inverse, this is true for $q = xy^{-1}$. Therefore $\mathbb{R}$ is a Euclidean ring.

Since we have established that the integers and the real numbers are Euclidean rings, any results we prove about Euclidean rings will, in particular, apply to these systems.

We will obtain another property of Euclidean rings.

THEOREM 12.1
Let R be a commutative ring and let x be any element of R. Then the set $I = \{x \cdot r \mid r \in R\}$ is an ideal of R.

Proof: We must show that (1) I is a subring of R and (2) for every $a \in R$ and $i \in I$, the products $a \cdot i$ and $i \cdot a$ are in I.

1. We will use the corollary to Theorem 11.4 to show that I is a subring. Choose elements $a = x \cdot r_1$ and $b = x \cdot r_2$ in I. Then

$$a - b = x \cdot r_1 - x \cdot r_2$$
$$= x(r_1 - r_2) \in I.$$

Also,

$$a \cdot b = (x \cdot r_1)(x \cdot r_2)$$
$$= x(r_1 \cdot x \cdot r_2) \in R.$$

Therefore I is a subring of R.

2. Choose elements $i = x \cdot r \in I$ and $a \in R$; then

$i \cdot a = (x \cdot r) \cdot a$
$ = x \cdot (r \cdot a) \in I.$

Also, since R is a commutative ring,

$a \cdot i = i \cdot a \in I$

by the above argument.

Therefore I is an ideal of R. ∎

DEFINITIONS
1. If R is a commutative ring and $x \in R$, then the ideal $\{x \cdot r \mid r \in R\}$ is called a **principal ideal** of R and will be denoted by (x).
2. If every ideal of an integral domain R is a principal ideal (x) for some $x \in R$, then R is called a **principal ideal domain.**

EXAMPLE 12.2
1. In the integers, (2) is the ideal of all even numbers.
2. In the real numbers, $(0) = \{0\}$ and for any $x \in \mathbb{R}$, $x \neq 0$, $(x) = \mathbb{R}$. (Why?)

THEOREM 12.2
Every Euclidean ring is a principal ideal domain.
Proof: Let I be any ideal of R, a Euclidean ring.
Case 1: If $I = \{0\}$, $(0) = \{0\}$ (Exercise 4).
Case 2: If $I \neq (0)$, we choose an element $i_o \in I$ such that $\delta(i_o) \leq \delta(i)$ for all $i \in I$. Note that this is always possible since $\delta(x)$ is a nonnegative integer for all $x \in I$.

Let x be any element of I. We wish to show that $x = i_o \cdot q$ for some $q \in R$. Since R is a Euclidean ring, there are elements q and r of R such that

$x = i_o \cdot q + r$

where $r = 0$ or $\delta(r) < \delta(i_o)$. We claim that r must equal 0.

We know that $x \in I$ and $i_o \in I$. Therefore $i_o \cdot q \in I$ since I is an ideal and thus,

$r = x - i_o \cdot q \in I.$

If $r \neq 0$, then $\delta(r) < \delta(i_o)$, which contradicts our choice of i_o. Therefore $r = 0$ and $x = i_o \cdot q$.

Thus $I = (i_o)$. That is, any ideal of R is a principal ideal and R is a principal ideal domain. ∎

In particular, this theorem says that any ideal of the integers consists of all multiples of some integer. That is, every ideal of Z is one of the sets $H_n = \{kn \mid k \in Z\}$.

One of the most useful properties of the integers is the notion of divisibility. This concept leads to many important and useful results such as the existence of a greatest common divisor, prime elements, unique factorization as a product of primes, and so on.

We will develop this material in the context of Euclidean rings; it will hold for the integers as well since the integers form a Euclidean ring.

DEFINITIONS
Let R be a Euclidean ring.

1. For elements $x, y \in R$, $x \neq 0$, we say that x divides y (written $x \mid y$) if there is a $z \in R$ such that $y = x \cdot z$. The notation $x \nmid y$ means that x does not divide y.

2. Let x and y be elements of R. If there is an element $d \in R$ such that
   a. $d \mid x$ and $d \mid y$ and
   b. if $z \in R$, $z \mid x$ and $z \mid y$, then $z \mid d$,

then d is called a **greatest common divisor** of x and y and is denoted $d = (x, y)$.

EXAMPLE 12.3
In the Euclidean ring of integers,
   1. $3 \mid 6$ since $6 = 3 \cdot 2$.
   2. A greatest common divisor of 6 and 9 is $(6, 9) = 3$ since $3 \mid 6$ and $3 \mid 9$ and the other divisors of 6 and 9, which are $1, -1,$ and -3, all divide 3. Note that -3 is also a greatest common divisor of 6 and 9 by our definition.

The next theorem guarantees the existence of greatest common divisors and gives us information about a method of finding it.

THEOREM 12.3
Let R be a Euclidean ring. If x and y are any nonzero elements of R, then there exists a greatest common divisor, (x, y), of the form $(x, y) = \alpha x + \beta y$ for some elements α and β of R.

Proof: Choose any two nonzero elements $x, y \in R$. Let $S_{x,y} = \{ax + by \mid a, b \in R\}$. We will first prove that $S_{x,y}$ is an ideal.

Choose s_1 and $s_2 \in S_{x,y}$. Then for some elements $a_1, b_1, a_2, b_2 \in R$, $s_1 = a_1 x + b_1 y$ and $s_2 = a_2 x + b_2 y$. Therefore

$$s_1 - s_2 = (a_1 - a_2)x + (b_1 - b_2)y \in S_{x,y}$$

and

$$s_1 \cdot s_2 = a_1 x a_2 x + a_1 x b_2 y + b_1 y a_2 x + b_1 y b_2 y$$
$$= (a_1 x a_2 + a_1 b_2 y) x + (b_1 a_2 x + b_1 y b_2) y \in S_{x,y}.$$

Therefore $S_{x,y}$ is a subring of R.

Also, for any $r \in R$,

$$r \cdot s_1 = (r a_1) x + (r a_2) y \in S_{x,y}$$

and

$$s_1 \cdot r = r \cdot s_1 \in S_{x,y}.$$

Therefore $S_{x,y}$ is an ideal of R.

By Theorem 12.2, $S_{x,y}$ is a principal ideal and therefore, for some $\alpha x + \beta y$, $S = (\alpha x + \beta y)$, or $\alpha x + \beta y$ divides every element of $S_{x,y}$.

We will show that $(x, y) = \alpha x + \beta y$.

$$x = 1 \cdot x + 0 \cdot y \in S_{x,y}$$

and

$$y = 0 \cdot x + 1 \cdot y \in S_{x,y}$$

so $(\alpha x + \beta y) | x$ and $(\alpha x + \beta y) | y$. Finally, let $z \in R$ be an element such that $z | x$ and $z | y$. Then $z | (ax + by)$ for any $a, b \in R$ (why?) and, in particular, $z | (\alpha x + \beta y)$. Therefore $(x, y) = \alpha x + \beta y$. ∎

EXERCISES

1. Show that $(Z, +, \cdot)$ is a Euclidean ring.
2. If R is any ring, $x \in R$, let $R_x = \{r \cdot x | r \in R\}$. Show that R_x is a left ideal of R.
3. Let R be any commutative ring. If there is a function $\delta : R - \{0\} \to Z^+$ that satisfies conditions 1 and 2 of the definition of a Euclidean ring, show that R must have an identity.
4. a. In any ring show that $(0) = \{0\}$.
   b. In $\mathbb{R}$, the ring of real numbers, show that if $x \in \mathbb{R}, x \neq 0$, then $(x) = \mathbb{R}$.

The next two exercises apply to any Euclidean ring R.

5. a. Prove that if $x | y$ and $y | z$, then $x | z$.
   b. Prove that if $x | y$ and $x | z$, then $x | (y + z)$ and $x | (y - z)$.
   c. Prove that if $x | y$, then $x | yz$ for any z in R.
6. a. Let d_1 and d_2 be greatest common divisors of x and y. Show that $d_1 | d_2$ and $d_2 | d_1$.
   b. How many greatest common divisors can a pair of integers have?
7. Find the greatest common divisors of the following pairs of integers:
   a. 13, -26 b. -48, 108
   c. 36, 78   d. 1, -1

8. In a field F:
   a. Show every nonzero element of F divides every other element of the field.
   b. For any nonzero elements $x, y, z \in F$, show that $z = (x, y)$.
   c. Prove that F is Euclidean.
9. Let u be an element of a Euclidean ring R such that for some $y \in R$, $u \cdot y = 1$, the multiplicative identity of R.
   a. Show that u divides every element of R.
   b. Show that the only such elements in the integers are $+1$ and -1.
10. Prove that the set of all u's defined in Exercise 9 forms a group under the multiplication of R.

In the last two exercises, we began exploring some properties of an important class of elements in a Euclidean ring.

DEFINITION

Let R be a Euclidean ring. If $u \in R$ has a multiplicative inverse in R, then we call u a **unit** in R.

That is, $u \in R$ is a unit if there is a $y \in R$ such that $u \cdot y = 1$, the multiplicative identity of R.

EXAMPLES 12.4
   1. 1 and -1 are the only units in the Euclidean ring of integers.
   2. In any field, any nonzero element is a unit.

DEFINITION

A nonunit element p in a Euclidean ring R is said to be a **prime** or **prime element** in R if $p = x \cdot y$ implies either x or y is a unit in R.

In other words, p is a prime if it cannot be properly factored.

EXAMPLE 12.5
In the Euclidean ring of integers, $2, 3, 5, 7, 11, \ldots$ are primes. Note that by this definition, $-2, -3, -5, -7, -11, \ldots$ are also primes.

We will prove two lemmas and then proceed to some major results on factorization.

LEMMA 1

In a Euclidean ring, if y is not a unit, then $\delta(x) < \delta(xy)$ for any $x \neq 0$.

Proof: Let $S_x = (x)$, the ideal of all multiples of x. By the proof of Theorem 12.2, $\delta(x) \leq \delta(z)$ for every $z \in S_x$. By definition, $\delta(x) \leq \delta(xy)$ for every y in R. If y is not a unit, we will prove $\delta(x) < \delta(xy)$ by contradiction.

Special classes of rings

Assume $\delta(x) = \delta(xy)$. By the second property in the definition of a Euclidean ring, there exist elements q and r in R such that

$$x = q(x \cdot y) + r \quad \text{where } r = 0 \quad \text{or} \quad \delta(r) < \delta(x \cdot y).$$

Again, since S_x is an ideal, $x \in S_x$, $q \cdot x \cdot y \in S_x$ and $x - q \cdot x \cdot y = r \in S_x$. If $r \neq 0$, $\delta(r) < \delta(x) = \delta(xy)$ which contradicts the choice of x. Thus $r = 0$ and $x = q \cdot x \cdot y = x(qy)$. By the cancellation laws in an integral domain, $qy = 1$. This contradicts the hypothesis that y is not a unit.

Therefore $\delta(x) < \delta(xy)$. ∎

LEMMA 2

In a Euclidean ring R, u is a unit if and only if $\delta(u) = \delta(1)$.

Proof: 1. Let u be a unit. We will show $\delta(u) = \delta(1)$. By hypothesis, $u \cdot v = 1$ for some $v \in R$. Then

$$\delta(u) \leq \delta(uv) = \delta(1).$$

But also,

$$\delta(1) \leq \delta(u \cdot 1) = \delta(u).$$

Therefore $\delta(u) \leq \delta(1)$ and $\delta(1) \leq \delta(u)$ implies $\delta(u) = \delta(1)$.

2. For the proof in the other direction, assume that $\delta(u) = \delta(1)$. Then there exist elements q and r in R such that

$$1 = q \cdot u + r \quad \text{where } r = 0 \quad \text{or} \quad \delta(r) < \delta(u) = \delta(1).$$

But the ideal $(1) = R$ (why?) and $\delta(1)$ is minimal in R so $\delta(r)$ cannot be less than $\delta(1)$. We conclude that $r = 0$.

Thus $1 = q \cdot u$ and u is a unit. ∎

THEOREM 12.4

If R is a Euclidean ring, every $x \in R$ is either a unit or a product of a finite number of primes of R.

Proof: Let X be the set of all elements in R that are not units and cannot be written as the product of a finite number of primes. We will prove that $X = \varnothing$ by contradiction.

If X is not empty, then $\{\delta(s) | s \in X\}$ is a nonempty subset of the nonnegative integers and therefore has a least element. Let x be the element of X for which $\delta(x)$ is minimal. Then we know

1. $\delta(x) > \delta(1)$ since x is not a unit and
2. if $r \in R$, with $\delta(r) < \delta(x)$, then either r is a unit or r is a product of a finite number of primes.

Since x is not prime, we can write

$$x = y \cdot z$$

where neither y nor z is a unit. By Lemma 1, we know that $\delta(y) < \delta(x)$ and $\delta(z) < \delta(x)$. Thus we know that y and z can each be written as a product of a finite number of primes, say

$$y = p_1 \cdot p_2 \cdot \cdots \cdot p_k$$
$$z = q_1 \cdot q_2 \cdot \cdots \cdot q_l.$$

But now $x = y \cdot z$, so

$$x = p_1 \cdot p_2 \cdot \cdots \cdot p_k \cdot q_1 \cdot q_2 \cdot \cdots \cdot q_l$$

which is a product of a finite number of primes. Therefore, $x \notin X$.

Since no x with minimal δ-value exists in X, X must be empty and every element of R is either a unit or a product of a finite number of primes. ∎

For the integers, this theorem is known as the Fundamental Theorem of Arithmetic.

EXERCISES

1. If $x|y$ and $y|x$ for elements x and y in a Euclidean ring, show that $y = u \cdot x$ for some unit u.
2. If $y = u \cdot x$ for some unit u, show that y divides x.
3. If $y = u \cdot x$ for some unit u, prove that $x = v \cdot y$ for some unit v.
4. Let $d_1 = (x, y)$ and $d_2 = (x, y)$. Show that $d_1 = u \cdot d_2$ for some unit u.
5. Let I be an ideal of R, a Euclidean ring. Show that if I contains a unit, then $I = R$.
6. If $x, y \in R$, a Euclidean ring, we say x and y are relatively prime if $(x, y) = 1$. Prove that $(x, y) = 1$ if $(x, y) = u$, for any unit u.
7. Let p be a prime and assume $(p, x) \neq 1$. Then show that $(p, x) = p \cdot u$ for a unit $u \in R$. (In particular, then $(p, x) = p \cdot 1 = p$.)
8. If p is a prime and $p \mid x$, show that $(p, x) = 1$.
9. If p is a prime and $p|(x_1 \cdot x_2)$, then prove that p divides at least one of the elements x_1 or x_2.
10. If p is a prime and $p|(x_1 \cdot x_2 \cdot \cdots \cdot x_n)$, then p divides at least one element of the set $\{x_1, x_2, \ldots, x_n\}$.
11. Show that if $x = u \cdot y$ and $y = v \cdot z$ where u and v are units, then $x = wz$ where w is a unit.
12. What are the prime elements of a field?

We turn now to a proof of the Unique Factorization Theorem for Euclidean rings. It is a theorem of fundamental importance, and is even of great interest when applied to the integers. We need one new concept.

Special classes of rings

DEFINITION

If $x = u \cdot y$ where u is a unit, then x and y are called **associates** in R.

EXAMPLE 12.6

In the integers, 3 and -3 are associates since $3 = (-1)(-3)$ for the unit (-1). Generalizing, we see that the only associates of an element $n \in Z$ are the elements $n = 1 \cdot n$ and $-n = (-1) \cdot n$. (Why?)

Note that Exercises 3 and 11 of the last set give us the results that:
1. If $x = u \cdot y$ for some unit u, then $y = v \cdot x$ for v a unit also.
2. If x and y are associates and y and z are associates, then x and z are associates.

THEOREM 12.5: UNIQUE FACTORIZATION THEOREM

Let R be a Euclidean ring and let $x \neq 0$ be any element of R which is not a unit. If

$$x = p_1 \cdot p_2 \cdot \cdots \cdot p_n = q_1 \cdot q_2 \cdot \cdots \cdot q_m$$

where p_i and q_j are primes for $i = 1, 2, \ldots, n$, $j = 1, 2, \ldots, m$, then $n = m$ and the primes can be renumbered, if necessary, so that p_i and q_i are associates.

Proof: Let

$$x = p_1 \cdot p_2 \cdot \cdots \cdot p_n = q_1 \cdot q_2 \cdot \cdots \cdot q_m$$

where the p's and q's are all primes and $m \geq n$. The prime p_1 divides $q_1 \cdot q_2 \cdot \cdots \cdot q_m$ and by Exercise 10 above, p_1 must divide at least one of $\{q_1, q_2, \ldots, q_m\}$. Renumber, if necessary, to say p_1 divides q_1. We can write $q_1 = u \cdot p_1$. Since p_1 and q_1 are primes, we know that u_1 must be a unit and p_1 and q_1 are associates.

Now we can write

$$x = p_1 \cdot p_2 \cdot \cdots \cdot p_n = u_1 \cdot p_1 \cdot q_2 \cdot \cdots \cdot q_m.$$

The cancellation laws in a Euclidean ring imply that

$$p_2 \cdot p_3 \cdot \cdots \cdot p_n = u_1 \cdot q_2 \cdot q_3 \cdot \cdots \cdot q_m.$$

In the last expression, we see that q_2 divides the product $p_2 \cdot p_3 \cdot \cdots \cdot p_n$ so that again by Exercise 10, q_2 divides one of the primes $p_2, p_3, \ldots, p_n$. Renumber, if necessary, to say that $q_2 | p_2$. Therefore $p_2 = v_2 \cdot q_2$ for some unit v_2, or equivalently, $q_2 = u_2 \cdot p_2$ for a unit u_2. p_2 and q_2 are associates. Our product can be rewritten as

$$p_2 \cdot p_3 \cdot \cdots \cdot p_n = u_1 \cdot u_2 \cdot p_2 \cdot q_3 \cdot \cdots \cdot q_n$$

and, omitting the common factor by the cancellation laws,

$$p_3 \cdot \cdots \cdot p_n = u_1 \cdot u_2 \cdot q_3 \cdot \cdots \cdot q_n.$$

We repeat this process, proving that p_i and q_i are associates and eliminating p_i and q_i at each step until we obtain

$$1 = u_1 \cdot u_2 \cdot \cdots \cdot u_n \cdot q_{n+1} \cdot \cdots \cdot q_m.$$

Since the primes $q_{n+1}, q_{n+2}, \ldots, q_m$ are not invertible, n must equal m and the proof is complete. ∎

This theorem says that there is essentially only one way to write an element in a Euclidean ring as a product of primes, except for the ambiguities resulting from multiplying by various units. In terms of the integers, the theorem says that if an integer is written as a product of prime numbers, $n = p_1 p_2 \cdot \cdots \cdot p_k$, then there is essentially no other way to factor n into a product of primes. For example,

$$18 = 2 \cdot 3 \cdot 3 = (-3) \cdot (-2) \cdot 3 = (-3) \cdot (2) \cdot (-3), \text{ etc.}$$

But any factorization of 18 as a product of primes contains one 2 and two 3's, up to multiplication by the units $+1$ and -1.

This theorem finishes our study of abstract Euclidean rings. The last topic in this chapter will be a discussion of polynomial rings. We will show that certain types of these rings are Euclidean rings which will then set the stage for the chapters on field theory.

EXERCISES

Let $Z[i] = \{n + mi \mid n, m \text{ are integers and } i = \sqrt{-1}\}$. $Z[i]$ is called the set of **Gaussian integers.**
1. Show that $Z[i]$ is an integral domain.
2. Let $\delta(n + mi) = n^2 + m^2$ and show $Z[i]$ is a Euclidean ring with this function δ. (*Hint:* $n^2 + m^2 = (n + mi)(n - mi)$.)
3. What are the units in $Z[i]$?
4. What are the primes in $Z[i]$?

POLYNOMIAL RINGS

Polynomial rings are a very important class of rings whose properties will be vital to our development of field theory. The specific examples of polynomials over the reals or the integers are familiar to most students; they are usually introduced in early high school.

We will introduce polynomials over a ring first and then specialize to consider polynomials over a field to get our major results.

DEFINITION

Let R be a ring. By a **polynomial over the ring R,** we mean a symbol of the form

$$a_0 + a_1 x + a_2 x^2 + \cdots + a_n x^n$$

where n is a nonnegative integer, $a_i \in R$, for $i = 0, 1, 2, \ldots, n$, and x is an indeterminate symbol.

$$R[x] = \{a_0 + a_1 x + a_2 x^2 + \cdots + a_n x^n \mid n \text{ is a nonnegative integer and } a_i \in R, i = 0, 1, 2, \ldots, n\}$$

is the **set of all polynomials over R.**

EXAMPLE 12.7

In the ring of real numbers $\mathbb{R}$, $-1 + 2x + 0x^3 - \frac{1}{2}x^4$, $\pi + 17x - 183x^2$, $3x - 4$, 5 and $\pi/4 x + x^{101}$ are examples of polynomials over the reals and elements of $\mathbb{R}[x]$.

DEFINITION

Let $a_0 + a_1 x + a_2 x^2 + \cdots + a_n x^n$ be a polynomial over a ring R where $a_n \neq 0$.
    1. The elements a_i, $i = 0, 1, \ldots, n$, are called the **coefficients** of the polynomial.
    2. The integer n is called the **degree** of the polynomial.

We will use the functional notation $p(x), q(x), r(x)$, and so on, to denote a polynomial.

EXAMPLE 12.8

In the polynomial $1 - x + 3x^2 - 2x^5$ over the ring of integers the coefficients are $a_0 = 1, a_1 = -1, a_2 = 3, a_3 = 0, a_4 = 0, a_5 = -2$. The degree of this polynomial is 5.

Note that for $0 \in R$, the polynomial

$$0 = 0 + 0x + 0x^2 + \cdots$$

has no degree.

DEFINITION

Two polynomials $p(x) = a_0 + a_1x + a_2x^2 + \cdots + a_nx^n$ and $q(x) = b_0 + b_1x + b_2x^2 + \cdots + b_nx^n$ are **equal** if $a_i = b_i$ for every $i = 0, 1, 2, \ldots, n$.

That is, we say $p(x) = q(x)$ if and only if their corresponding coefficients are equal.

We will define operations of addition and multiplication on $R[x]$, the set of all polynomials over the ring R. These definitions must be consistent with the addition and multiplication in the ring R since for an element $a \in R$, the symbol $p(x) = a$ is a polynomial of degree zero. We would also like these operations to coincide with the addition and multiplication of polynomials that has been taught in the lower level algebra courses. That is, we want

$$(4 - 3x + x^2) + (5 + x - 5x^2 + x^3) = 9 - 2x - 4x^2 + x^3$$

and

$$(3 + x)(4 - x) = 12 + x - x^2.$$

We make the following definitions.

DEFINITION

Let $R[x]$ be the set of all polynomials over a ring R. For $p(x)$ and $q(x) \in R[x]$,

$$p(x) = a_0 + a_1x + \cdots + a_nx^n$$
$$q(x) = b_0 + b_1x + \cdots + b_mx^m$$

where $m \geq n$, we define the **sum** of $p(x)$ and $q(x)$ by

$$p(x) + q(x) = (a_0 + b_0) + (a_1 + b_1)x + \cdots + (a_n + b_n)x^n$$
$$+ (0 + b_{n+1})x^{n+1} + \cdots + (0 + b_m)x^m.$$

EXAMPLES 12.9

1. In $\mathbb{R}[x]$, the set of polynomials over the reals, the sum of $p(x) = -1 + 3x - 17x^2$ and $q(x) = 2 - \frac{5}{2}x + \frac{7}{3}x^2 - x^3$ is

$$p(x) + q(x) = 1 + \tfrac{1}{2}x - \tfrac{44}{3}x^2 - x^3.$$

2. In $Z_6[x]$, the set of polynomials over the ring Z_6, the sum of $p(x) = 5 + 3x + 3x^2 - 3x^3$ and $q(x) = 2 + 5x - 2x^2 - 3x^3$ is

$$p(x) + q(x) = (5 + 2) + (3 + 5)x + (3 - 2)x^2 + (-3 - 3)x^3$$
$$= 1 + 2x + x^2.$$

Special classes of rings 173

Note that the coefficients are combined by the addition in the underlying ring ($5 + 2 = 1$ in Z_6, and so on).

In defining multiplication, we want to preserve the identity $x^k x^l = x^{k+l}$.

DEFINITION

In $R[x]$, the set of all polynomials over a ring R, the **product** of
$$p(x) = a_0 + a_1 x + a_2 x^2 + \cdots + a_n x^n$$
and
$$q(x) = b_0 + b_1 x + b_2 x^2 + \cdots + b_m x^m$$
is the polynomial
$$p(x) \cdot q(x) = c_0 + c_1 x + c_2 x^2 + \cdots + c_{n+m} x^{n+m}$$
where
$$c_k = a_k b_0 + a_{k-1} b_1 + \cdots + a_{k-r} b_r + \cdots + a_0 b_k$$
for $k = 0, 1, \ldots, n + m$.

EXAMPLE 12.10

Let us use this definition to compute the product of the polynomials
$$p(x) = 3 - x + 2x^2 = a_0 + a_1 x + a_2 x^2$$
and
$$q(x) = 5 + 2x - 3x^2 + x^3 = b_0 + b_1 x + b_2 x^2 + b_3 x^3.$$

$c_0 = a_0 b_0 = 5(3) = 15$
$c_1 = a_1 b_0 + a_0 b_1 = (-1) \cdot (5) + (3)(2) = 1$
$c_2 = a_2 b_0 + a_1 b_1 + a_0 b_2 = 2 \cdot 5 + (-1) \cdot 2 + 3 \cdot (-3) = -1$
$c_3 = a_3 b_0 + a_2 b_1 + a_1 b_2 + a_0 b_3 = 10$
$c_4 = a_4 b_0 + a_3 b_1 + a_2 b_2 + a_1 b_3 + a_0 b_4 = -7$
$c_5 = a_5 b_0 + a_4 b_1 + a_3 b_2 + a_2 b_3 + a_1 b_4 + a_0 b_5 = 2$
$c_6 = a_6 b_0 + a_5 b_1 + \cdots + a_0 b_6 = 0.$

We note that $c_k = 0$ for $k \geq 6$. Therefore
$$p(x) \cdot q(x) = c_0 + c_1 x + c_2 x^2 + \cdots + c_5 x^5$$
$$= 15 + x - x^2 + 10x^3 - 7x^4 + 2x^5.$$

Notation: We will use the summation $p(x) = \sum_{k=0}^{n} a_k x^k$ to denote the polynomial $p(x) = a_0 + a_1 x + a_2 x^2 + \cdots + a_n x^n$. It makes the calculations necessary to prove the next theorem considerably easier.

The addition of the polynomials $p(x) = \sum_{k=0}^{n} a_k x^k$ and $q(x) = \sum_{k=0}^{m} b_k x^k$ reads

$$p(x) + q(x) = \sum_{k=0}^{\max(n,\, m)} (a_k + b_k) x^k.$$

The product is

$$p(x) \cdot q(x) = \sum_{k=0}^{n+m} \left(\sum_{i=0}^{k} a_{k-i} b_i \right) x^k.$$

When adding polynomials $p(x)$ and $q(x)$ of degree n and m respectively, we will consider all coefficients $(a_i + b_i)$ of the sum up to the larger of the integers n and m. The undefined coefficients of the polynomial of smaller degree will be taken as zero.

THEOREM 12.6

If R is a ring, $R[x]$ is a ring.

Proof: We need to verify all the ring axioms. Most of them will be done here, but a few will be left as exercises.

1. We need to show that $R[x]$ is an abelian group under the operation $+$.
   a. The set $R[x]$ is closed under $+$ by the definition of the operation.
   b. To prove the associative law of addition, we choose elements $p(x) = \sum_{k=0}^{n} a_k x^k$, $q(x) = \sum_{k=0}^{n} b_k x^k$, $r(x) = \sum_{k=0}^{n} c_k x^k$ in $R[x]$. Then

$$[p(x) + q(x)] + r(x) = \sum_{k=0}^{n} (a_k + b_k) x^k + \sum_{k=0}^{n} c_k x^k$$

$$= \sum_{k=0}^{n} [(a_k + b_k) + c_k] x^k$$

$$= \sum_{k=0}^{n} [a_k + (b_k + c_k)] x^k \quad \text{(since } R \text{ is associative)}$$

$$= \sum_{k=0}^{n} a_k x^k + \sum_{k=0}^{n} (b_k + c_k) x^k$$

$$= p(x) + [q(x) + r(x)].$$

   c. The polynomial $0 = 0 + 0x + 0x^2 + \cdots$ acts as an identity since $p(x) + 0 = 0 + p(x) = p(x)$.

d. The additive inverse of the polynomial $p(x) = \sum_{k=0}^{n} a_k x^k$ is the polynomial $-p(x) = \sum_{k=0}^{n} (-a_k) x^k$ since

$$p(x) + [-p(x)] = \sum_{k=0}^{n} a_k x^k + \sum_{k=0}^{n} (-a_k) x^k$$

$$= \sum_{k=0}^{n} (a_k + (-a_k)) x^k$$

$$= \sum_{k=0}^{n} 0 x^k = 0.$$

e. $R[x]$ is abelian since for

$$p(x) = \sum_{k=0}^{n} a_k x^k \quad \text{and} \quad q(x) = \sum_{k=0}^{n} b_k x^k \quad \text{in } R[x]$$

$$p(x) + q(x) = \sum_{k=0}^{n} (a_k + b_k) x^k$$

$$= \sum_{k=0}^{n} (b_k + a_k) x^k \quad \text{(since R is abelian)}$$

$$= q(x) + p(x).$$

2. $R[x]$ is closed under $\cdot$ by the definition of the operation.

3. Showing that $R[x]$ is associative is straightforward and is left as an exercise.

4. To show that $R[x]$ satisfies the distributive laws, choose polynomials $p(x) = \sum_{k=0}^{m} a_k x^k$, $q(x) = \sum_{k=0}^{n} b_k x^k$, and $r(x) = \sum_{k=0}^{n} c_k x^k$. Then

$$p(x)[q(x) + r(x)] = \left(\sum_{k=0}^{m} a_k x^k\right)\left(\sum_{k=0}^{n} (b_k + c_k) x^k\right)$$

$$= \sum_{k=0}^{n+m} \left(\sum_{i=0}^{k} [a_{k-i}(b_i + c_i)]\right) x^k$$

$$= \sum_{k=0}^{n+m} \left(\sum_{i=0}^{k} (a_{k-i} b_i + a_{k-i} c_i)\right) x^k \quad \text{(why?)}$$

$$= \sum_{k=0}^{n+m} \left(\sum_{i=0}^{k} a_{k-i} b_i\right) x^k + \sum_{k=0}^{n+m} \left(\sum_{i=0}^{k} a_{k-i} c_i\right) x^k$$

$$= p(x)q(x) + p(x)r(x).$$

The other distributive law is proved similarly and is left as an exercise. ■

We will investigate some of the properties of a ring R which are inherited by the polynomial ring $R[x]$.

THEOREM 12.7
If R is a commutative ring, then $R[x]$ is commutative.
    Proof: Let $p(x) = \sum_{k=0}^{n} a_k x^k$ and $q(x) = \sum_{k=0}^{m} b_k x^k$ be elements of $R[x]$. Then

$$p(x) \cdot q(x) = \sum_{k=0}^{n+m} \left(\sum_{j=0}^{k} a_j b_{k-j} \right) x^k$$

$$= \sum_{k=0}^{n+m} \left(\sum_{j=0}^{k} b_{k-j} a_j \right) x^k \quad \text{(since } R \text{ is commutative)}$$

$$= \sum_{k=0}^{n+m} \left(\sum_{l=0}^{k} b_l a_{k-l} \right) x^k \quad \text{(by the change of index } l = k - j\text{)}$$

$$= q(x) \cdot p(x). \quad \blacksquare$$

THEOREM 12.8
If R is a ring with multiplicative identity, then $R[x]$ has a multiplicative identity also.
    Proof: Let 1 be the multiplicative identity of R. Then $1 = 1 + 0x + 0x^2 + \cdots$ also acts as an identity for $R[x]$. The details are left as an exercise. $\blacksquare$

THEOREM 12.9
If R is an integral domain, then $R[x]$ is also an integral domain.
    Proof: That $R[x]$ is a commutative ring with identity was proved in Theorems 12.6, 12.7, and 12.8. All we have to check is that if $p(x)$ and $q(x)$ are any nonzero polynomials, then $p(x) \cdot q(x) \neq 0$.
    Let

$$p(x) = \sum_{k=0}^{n} a_k x^k \quad \text{and} \quad q(x) = \sum_{k=0}^{m} b_k x^k$$

where $a_n \neq 0$ and $b_m \neq 0$. Then the term of degree $n + m$ in $p(x) \cdot q(x)$ has coefficient $a_n b_m$ which is not zero since R is an integral domain. Thus $p(x)q(x) \neq 0$. $\blacksquare$

EXERCISES
1. Perform the following sums:
    a. $(-13 + 73x^2 + 104x^5) + (8 + 2x - 3\frac{1}{2}x^2 + 35x^4)$
    b. $(4) + (96x - 2x^3)$
    c. $(-3 \cdot 25 - \pi x + \pi^2 x^2) + (x^{107} - x^{113})$.

2. Perform the following products:
   a. $(-2 + 3x - x^2 + 4x^3) \cdot (x + 4x^3 - 2x^4)$
   b. $(-1 + x + x^2) \cdot (-3 + x^7)$
   c. $(1 - x) \cdot (1 + x + x^2 + x^3 + x^4 + x^5)$.
3. Show that multiplication of polynomials is associative.
4. If R is a ring, show that $(p(x) + q(x)) \cdot r(x) = p(x)r(x) + q(x) \cdot r(x)$ for any $p(x)$, $q(x)$, and $r(x)$ in $R[x]$.
5. Let $p(x)$ be a polynomial of degree n and $q(x)$ be a polynomial of degree m. Show that the degree of $p(x) + q(x)$ is less than or equal to the larger of n and m.
6. a. If $p(x)$ and $q(x)$ are as in Exercise 5, show that the degree of $p(x) \cdot q(x)$ is less than or equal to $n + m$.
   b. Under what conditions does equality hold?
7. Complete the proof that R has an identity implies $R[x]$ has an identity.
8. Show that R is a subring of $R[x]$.
9. a. If F is a field, what are the units in $F[x]$?
   b. Is $F[x]$ a field? Prove your result.

Before we continue, let us notice the results obtained in the exercises about the degree of a polynomial. If $p(x)$ is a polynomial of degree n and $q(x)$ is a polynomial of degree m in $R[x]$, the ring of polynomials over a ring R, then

1. the degree of $p(x) + q(x) \leq \max(n, m)$.
2. the degree of $p(x) \cdot q(x) \leq n + m$.
3. the degree of $p(x) \cdot q(x) = n + m$ when R is an integral domain.

We want to investigate polynomial rings which are Euclidean rings. The fact that R is a Euclidean ring is not enough to guarantee that $R[x]$ is a Euclidean ring. Therefore we will be considering a field F and the ring $F[x]$ of polynomials over F.

THEOREM 12.10

If F is a field, $F[x]$ is a Euclidean ring.

Proof: We need to find a function $\delta: F[x] \to Z^+$ with the property that for every nonzero $p(x)$ and $q(x)$ in $F[x]$,

1. $\delta(p(x)) \leq \delta(p(x)q(x))$ and
2. there are polynomials $t(x)$ and $r(x)$ in $F[x]$ such that

$p(x) = t(x)q(x) + r(x)$ where $r(x) = 0$

or

$\delta(r(x)) < \delta(q(x))$.

Define $\delta(p(x))$ = the degree of $p(x)$.
1. First, by property 3 above, we know that $\delta(p(x) \cdot q(x)) = \delta(p(x)) + \delta(q(x))$. Thus if $p(x), q(x) \neq 0$, $\delta(q(x)) \geq 0$ and $\delta(p(x)) \leq \delta(p(x) \cdot q(x)) = \delta(p(x)) + \delta(q(x))$.

2. *Case 1:* If $\delta(p(x)) < \delta(q(x))$, take $t(x) = 0$ and $r(x) = p(x)$. Then

$$p(x) = q(x) \cdot 0 + p(x) \quad \text{and} \quad \delta(p(x)) < \delta(q(x))$$

Case 2: Assume $\delta(q(x)) \leq \delta(p(x))$. We will prove the result by contradiction.

Let $X =$ the set of all polynomials $p(x)$ for which there is a polynomial $q(x)$ with $\delta(q(x)) < \delta(p(x))$ and for which there are no polynomials $t(x)$ and $r(x)$ with the desired property.

If $X \neq \emptyset$, then $\{\delta(p(x)) | p(x) \in X\}$ is a nonempty subset of the nonnegative integers and therefore has a least element. Let $p_0(x) = \sum_{k=0}^{m} a_k x^k$ be that polynomial of minimal degree in X and let $q_0 = \sum_{k=0}^{n} b_k x^k$ with $n \leq m$ be the other polynomial of the pair for which the theorem fails.

We will show that X cannot be nonempty by finding polynomials $t_0(x)$ and $r_0(x)$ that satisfy the theorem.

Let $p(x) = p_0(x) - a_m b_n^{-1} x^{m-n} q_0(x)$. The polynomial $p(x)$ has $\delta(p(x)) < \delta(p_0(x)) = m$ and therefore there are polynomials $t(x)$ and $r(x)$ such that

$$p(x) = t(x) q_0(x) + r(x)$$

where

$$r(x) = 0 \quad \text{or} \quad \delta(r(x)) < \delta(q_0(x)).$$

Finally, we have

$$\begin{aligned} p_0(x) &= p(x) + a_m b_n^{-1} x^{m-n} q_0(x) \\ &= t(x) q_0(x) + a_m b_n^{-1} x^{m-n} q_0(x) + r(x) \\ &= (t(x) + a_m b_n^{-1} x^{m-n}) q_0(x) + r(x). \end{aligned}$$

Let $t_0(x) = t(x) + a_m b_n^{-1} x^{m-n}$ and let $r_0(x) = r(x)$. Note that either $r_0(x) = 0$ or $\delta(r_0(x)) = \delta(r(x)) < \delta(q_0(x))$.

Since we cannot find an element of X with least degree, $X = \emptyset$ and the theorem is proved. ∎

COROLLARY

If F is a field, then $F[x]$ is a principal ideal domain.

Proof: Every Euclidean ring is a principal ideal domain. ∎

We conclude our study of polynomial rings with a result which will be useful in the next part of the book.

First, notice that if F is a field, the units of $F[x]$ are precisely the polynomials of degree zero. That is, if $u(x) = a$ for $a \in F$, $a \neq 0$, then $[u(x)]^{-1} = a^{-1}$. Remember that the polynomial $0 = 0 + 0x + 0x^2 + \cdots$ has no degree.

Special classes of rings

The definition of a prime element can be applied to the ring of all polynomials $F[x]$ over a field F. A polynomial $p(x)$ is a prime element of $F[x]$ if whenever $p(x) = a(x) \cdot b(x)$, either $a(x)$ or $b(x)$ must be a unit, a polynomial of degree zero.

DEFINITION

A prime polynomial is said to be an **irreducible** polynomial.
If $p(x)$ is not irreducible, it is called a **reducible** polynomial.

EXAMPLE 12.11

If $\mathbb{R}$ is the set of real numbers, then in $\mathbb{R}[x]$, the polynomial $2x + 2$ is prime since only nonzero constants (units) can be factored out.
$(x^2 + x + 3)$ is also prime, but $x^2 - 2 = (x + \sqrt{2})(x - \sqrt{2})$ is not prime.

THEOREM 12.11

Let $F[x]$ be a polynomial ring over F, a field. Let $M = (p(x))$ be the ideal of $F[x]$ generated by the element $p(x)$. Then $F[x]|M$ is a field if and only if $p(x)$ is irreducible.

Proof: 1. If $p(x)$ is reducible, then we can write

$$p(x) = a(x) \cdot b(x)$$

where neither $a(x)$ nor $b(x)$ is a unit. Thus $0 < \text{degree } a(x) < \text{degree } p(x)$ and $0 < \text{degree } b(x) < \text{degree } p(x)$ so that the coset $M + a(x) \neq M$ and $M + b(x) \neq M$, where M is the additive identity (the zero) of $F[x]|M$. However,

$$(M + a(x)) \cdot (M + b(x)) = M + a(x) \cdot b(x)$$
$$= M + p(x)$$
$$= M \quad \text{(the zero in } F[x]|M\text{)}$$

so that $F[x]|M$ is not even an integral domain if $p(x)$ is reducible.

2. Next, assume that $p(x)$ is irreducible. We will prove $M = (p(x))$ is a maximal ideal. That is, we will show that there is no ideal $(q(x))$ such that

$$(p(x)) \subseteq (q(x)) \subseteq F[x]$$

where $(q(x))$ is not equal to either $(p(x))$ or $F[x]$.

By contradiction, assume there is an ideal $(q(x))$ strictly between $F[x]$ and $(p(x))$. Then we have $(p(x)) \subseteq (q(x)) \subseteq F[x]$. Thus $p(x) \in (q(x))$, so that $p(x) = a(x) \cdot q(x)$ for some $a(x) \in F[x]$. But $p(x)$ is irreducible, so either $a(x)$ or $q(x)$ has degree 0.

If $q(x)$ has degree 0, then $(q(x)) = F[x]$. (Why?)

If $a(x)$ has degree 0, then $q(x) = a(x)^{-1}p(x)$ and the ideals $(q(x)) = (p(x))$. (Why?)

Therefore we have shown that $(p(x))$ is a maximal ideal of $F[x]$ if $p(x)$ is irreducible.

Finally, to show that $F[x]/M$ is a field, it suffices to show that every element that is not equal to M, the zero in $F[x]/M$, has a multiplicative inverse relative to the multiplicative identity $M + 1$.

Choose a polynomial $a(x)$ that is not in $M = (p(x))$. Let

$$S = \{\alpha(x) \cdot a(x) + \beta(x) \cdot p(x) | \alpha(x), \beta(x) \in F[x]\}.$$

We claim that S is an ideal of $F[x]$. (This part of the proof is left as an exercise.)

Also, $(p(x)) \subseteq S$ and $a(x) \in S$. Therefore, since $a(x) \notin (p(x))$, and the ideal $(p(x))$ is maximal, we may conclude that $S = F[x]$. But then, for some particular $\alpha(x), \beta(x) \in S$, we must have

$$\alpha(x) \cdot a(x) + \beta(x) \cdot p(x) = 1$$

which implies that the product of the cosets

$$(M + \alpha(x)) \cdot (M + a(x)) = M + \alpha(x)a(x)$$
$$= M + 1$$

and therefore $M + a(x)$ has a multiplicative inverse in $F[x]/M$, which completes the proof. ∎

This concludes this part of the book on rings. We now have enough information to begin our study of fields and their properties.

EXERCISES

1. Which of the following are irreducible over (a) the real numbers, (b) the rational numbers, (c) the complex numbers:
    (i) $x^2 - 4$
    (ii) $x^3 + 4$
    (iii) $x^2 - 1$
    (iv) $x^2 + x + 1$
    (v) $x^4 + 1$
    (vi) $x^2 + 2x + 1$.
2. Let F be a field. Show that if the characteristic of F is nonzero then there is a subfield (a subset of F which is a field under the same operations) of F containing 0 and 1 whose order is equal to the characteristic of F.
3. Prove that for elements $a(x), p(x) \in F[x]$, the set $S = \{\alpha(x) \cdot a(x) + \beta(x) \cdot p(x) | \alpha(x), \beta(x) \in F[x]\}$ is an ideal of $F[x]$.
4. Write out addition and multiplication tables for $(Z_5, +, \cdot)$, a field of order 5. Find three polynomials of degree >1 which are irreducible over F_5.
5. If $p(x) = x^2 + 1 \in \mathbb{R}[x]$:
    a. Find $(p(x))$.
    b. Find $\mathbb{R}[x]/(p(x))$.

FIELD THEORY

13

Examples and axioms

In this last part of the book, we study fields and some of their properties. We have already defined fields and studied polynomial rings over fields. This investigation will allow us to study fields and their extensions in greater detail. We will also be looking at some classical problems that can be discussed in the context of field theory.

DEFINITION
A **field** is a ring $(F, +, \cdot)$ with the additional property that $(F - \{0\}, \cdot)$ is an abelian group.

A field has two group structures: $(F, +)$ is an abelian group and $(F - \{0\}, \cdot)$ is also an abelian group. The concepts that were developed for groups in Part I will be used in this part also.

A field is a special ring, an integral domain, and also a Euclidean ring. (These facts are left as exercises.) The material in Part II is valid for fields and we will be using it also.

Let us look at some examples of fields with which we are already familiar.

EXAMPLES 13.1

1. The system of real numbers $(\mathbb{R}, +, \cdot)$ forms a field. This will serve as one of our most important examples since its familiarity makes it very useful in building intuition and understanding examples.

2. The set of all complex numbers
$$\mathbb{C} = \{a + bi \,|\, a, b \in \mathbb{R} \text{ and } i = \sqrt{-1}\}$$

is a field with operations + and · defined by

$$(a + bi) + (c + di) = (a + c) + (b + d)i$$

and

$$(a + bi) \cdot (c + di) = (ac - bd) + (ad + bc)i.$$

The multiplicative identity of $\mathbb{C}$ is $1 + 0i = 1$ and the multiplicative inverse of $a + bi \neq 0$ is $\dfrac{a}{a^2 + b^2} - \dfrac{b}{a^2 + b^2} i$. For each real number a, $a = a + 0i \in \mathbb{C}$ so that the reals are a subset of the complex numbers.

3. For p, a prime integer, consider the ring $Z_p = \{0, 1, \ldots, p - 1\}$ under addition and multiplication modulo p. Z_p is a commutative ring with identity. We will show that Z_p is (a) an integral domain and (b) a field.

a. To show that Z_p is an integral domain, we need to show that Z_p has no zero divisors. That is, we need to show that $x \cdot y = 0$ implies $x = 0$ or $y = 0$.

If $x, y \in Z_p$ and $x \cdot y \equiv 0 \pmod{p}$, then $x \cdot y = kp$ for some integer k. The prime p must divide either x or y. The only element in $Z_p = \{0, 1, 2, \ldots, p - 1\}$ that is divisible by p is 0 so that either x or y must be zero. Therefore Z_p is an integral domain.

b. To show that Z_p is a field, we need to show that any nonzero element in Z_p has a multiplicative inverse. Choose $x \in Z_p$, $x \neq 0$, and consider the products

$$x \cdot 1, x \cdot 2, \ldots, x \cdot (p - 1).$$

By part a and the cancellation laws, these are all distinct nonzero elements of Z_p and therefore for some y, $1 \leq y \leq p - 1$, $x \cdot y = 1$ and $y = x^{-1}$.

In this chapter we also wish to develop another source for examples of a field. Specifically, we will show that every integral domain can be considered to be a subset of a field, or that a field can be constructed from any integral domain. The field of rational numbers can be constructed from the integers in this way. We will do the construction generally for any integral domain D.

Let D be an integral domain and let

$$E = \left\{ \frac{a}{b} \,\Big|\, a, b \in D, b \neq 0 \right\}$$

be the set of formal quotients of elements of D. Note that the symbol a/b

Examples and axioms

has no meaning in D. We will give it a meaning in E by defining an equality of these elements and two operations on them.

DEFINITION
If a/b and c/d are elements of E, we say that $a/b = c/d$ when $a \cdot d = b \cdot c$ where $\cdot$ is the multiplication in D.

Note: From this point on we will omit the $\cdot$ and refer to the product of elements x and y in D by xy.

THEOREM 13.1
The relation "$a/b = c/d$ when $ad = bc$" is an equivalence relation on E.

Proof: We must show that this relation is (1) reflexive, (2) symmetric, and (3) transitive.

1. For every $a/b \in E$, $a/b = a/b$ since $ab = ba$ and the relation is reflexive.
2. If $a/b = c/d$, then $ad = bc$ which is the same as $cb = da$ by the commutativity of D. Therefore $c/d = a/b$ and the relation is symmetric.
3. If $a/b = c/d$ and $c/d = e/f$, we have the equalities $ad = bc$ and $cf = de$. Multiplying the first equality by f and the second by b, we get $bcf = adf$ and $bcf = bde$. Therefore

$$adf = bde$$

and the cancellation laws give

$$af = be$$

or

$$\frac{a}{b} = \frac{e}{f}.$$

Since this relation satisfies all three of the properties, it is an equivalence relation. ∎

We recall that the equivalence classes induced by an equivalence relation form a partition of the underlying set E. An equivalence class is a set of the form

$$\left(\frac{a}{b}\right) = \left\{\frac{c}{d} \middle| \frac{a}{b} = \frac{c}{d}; a, b, c, d \in D; b, d \neq 0\right\}.$$

Any two equivalence classes are either disjoint or equal. Every element of E belongs to one equivalence class and any element of an equivalence class can be taken as the representative of its class.

Let F be the set of all equivalence classes for this relation:

$$F = \left\{ \left(\frac{a}{b}\right) \middle| \frac{a}{b} \in E \right\}.$$

EXAMPLE 13.2
If we take the set of all integers Z as the integral domain, then the set of all formal quotients of integers with nonzero denominators is the set of all rational numbers. The equivalence relation guarantees that all fractions like $\frac{1}{2}, \frac{4}{8}, \frac{7}{14}, \frac{-13}{-26}$, and so on, are taken to be the same rational number.

We will define two operations, $+$ and $\cdot$, on F so that the system $(F, +, \cdot)$ is a field.

Choose elements (a/b) and (c/d) of F. Then a/b and c/d are elements of E which guarantees that $b \neq 0$ and $d \neq 0$. Since D is an integral domain, $bd \neq 0$ and the elements $ad + bc/bd$ and ac/bd belong to E while their equivalence classes $(ad + bc/bd)$ and (ac/bd) are elements of F.

DEFINITION
For two elements (a/b) and (c/d) in F, we define

$$\left(\frac{a}{b}\right) + \left(\frac{c}{d}\right) = \left(\frac{ad + bc}{bd}\right)$$

and

$$\left(\frac{a}{b}\right) \cdot \left(\frac{c}{d}\right) = \left(\frac{ac}{bd}\right)$$

where the product and the sum within each quotient are those of the integral domain D.

Note: Since the operations are defined in terms of a particular representative of an equivalence class, it must be established that the computations of both operations do not depend on the choice of representative from that class. These operations turn out to be the usual operations of addition and multiplication on the set of rational numbers.

THEOREM 13.2
The operations $+$ and $\cdot$ defined on F, the set of all equivalence classes of formal quotients, by

$$\left(\frac{a}{b}\right) + \left(\frac{c}{d}\right) = \left(\frac{ad + bc}{bd}\right)$$

and

$$\left(\frac{a}{b}\right) \cdot \left(\frac{c}{d}\right) = \left(\frac{ac}{bd}\right)$$

are well defined.

Proof: Choose elements $a_1/b_1 \in (a/b)$ and $c_1/d_1 \in (c/d)$. Then $a_1/b_1 = a/b$ and $c_1/d_1 = c/d$ or

$$a_1 b = a b_1 \quad \text{and} \quad c_1 d = c d_1.$$

The sum

$$\left(\frac{a_1}{b_1}\right) + \left(\frac{c_1}{d_1}\right) = \left(\frac{a_1 d_1 + b_1 c_1}{b_1 d_1}\right).$$

We need to show that

$$\left(\frac{a_1 d_1 + b_1 c_1}{b_1 d_1}\right) = \left(\frac{ad + bc}{bd}\right).$$

Since $a_1 b = a b_1$,

$$a_1 b d d_1 = a b_1 d d_1.$$

Also since $c_1 d = c d_1$,

$$c_1 d b_1 b = c d_1 b_1 b.$$

Rearranging and adding, we get

$$a_1 d_1 bd + b_1 c_1 bd = adb_1 d_1 + bcb_1 d_1$$

or

$$(a_1 d_1 + b_1 c_1) bd = b_1 d_1 (ad + bc)$$

which is equivalent to

$$\frac{a_1 d_1 + b_1 c_1}{b_1 d_1} = \frac{ad + bc}{bd}.$$

Since the elements are equal, so are their equivalence classes and addition is well defined.

The proof that multiplication is well defined is left as an exercise. ∎

THEOREM 13.3

The system $(F, +, \cdot)$ is a field.

Proof: We need to show that (1) $(F, +)$ is an abelian group; (2) $(F - \{0\}, \cdot)$ is an abelian group; and (3) the distributive laws hold.

1. a. The operation $+$ on F is well defined by the last theorem.
  b. Choose elements (a/b), (c/d), and (e/f) in F. The sum

$$\left[\left(\frac{a}{b}\right) + \left(\frac{c}{d}\right)\right] + \left(\frac{e}{f}\right) = \left(\frac{ad + bc}{bd}\right) + \left(\frac{e}{f}\right)$$

$$= \left(\frac{(ad + bc)f + ebd}{bdf}\right)$$

$$= \left(\frac{adf + bcf + ebd}{bdf}\right)$$

while

$$\left(\frac{a}{b}\right) + \left[\left(\frac{c}{d}\right) + \left(\frac{e}{f}\right)\right] = \left(\frac{a}{b}\right) + \left(\frac{cf + de}{df}\right)$$

$$= \left(\frac{adf + b(cf + de)}{bdf}\right)$$

$$= \left(\frac{adf + bcf + ebd}{bdf}\right).$$

Since both sums have the same value, $+$ is associative.

  c. The operation $+$ is commutative since

$$\left(\frac{a}{b}\right) + \left(\frac{c}{d}\right) = \left(\frac{ad + bc}{bd}\right)$$

$$= \left(\frac{cb + da}{db}\right)$$

$$= \left(\frac{c}{d}\right) + \left(\frac{a}{b}\right).$$

  d. The element $(0/1)$ acts as an identity for $(F, +)$ since

$$\left(\frac{a}{b}\right) + \left(\frac{0}{1}\right) = \left(\frac{a \cdot 1 + b \cdot 0}{b \cdot 1}\right) = \left(\frac{a}{b}\right).$$

  e. The element $(-a/b)$ is an additive inverse to (a/b) since

$$\left(\frac{a}{b}\right) + \left(\frac{-a}{b}\right) = \left(\frac{ab + (-a)b}{bb}\right) = \left(\frac{0}{b^2}\right) = \left(\frac{0}{1}\right). \quad \text{(Why?)}$$

2. The proof that $(F - \{(0/1)\}, \cdot)$ is an abelian group is left as an exercise.

Examples and axioms 189

3. Choose (a/b), (c/d), and (e/f) in F. Then

$$\left(\frac{a}{b}\right) \cdot \left[\left(\frac{c}{d}\right) + \left(\frac{e}{f}\right)\right] = \left(\frac{a}{b}\right) \cdot \left(\frac{cf + ed}{df}\right)$$

$$= \left(\frac{acf + aed}{bdf}\right)$$

while

$$\left(\frac{a}{b}\right)\left(\frac{c}{d}\right) + \left(\frac{a}{b}\right)\left(\frac{e}{f}\right) = \left(\frac{ac}{bd}\right) + \left(\frac{ae}{bf}\right)$$

$$= \left(\frac{acbf + aebd}{bdbf}\right)$$

and we see that

$$\left(\frac{acbf + aebd}{bdbf}\right) = \left(\frac{acf + aed}{bdf}\right). \qquad \text{(Why?)}$$

Verifying the other distributive law should be done as an exercise. ∎

DEFINITION
F is called the **field of quotients** of D.

EXAMPLE 13.3
Again, note that this construction will yield the rationals if our initial integral domain is the integers. The field of rational numbers is therefore the field of quotients of the integers and each rational number can be thought of as an equivalence class of quotients.

It is elementary that the integers are a subset of the rationals but we have to prove, in general, that every integral domain is, in some sense, contained in its field of quotients. Let

$$S = \left\{\left(\frac{a}{1}\right) \middle| a \in D \text{ and 1 is the multiplicative identity in } D\right\}.$$

We will show that the set S is isomorphic to the set D. This will complete our discussion of integral domains and their fields of quotients.

THEOREM 13.4
The function $\alpha: D \to S$ defined by

$$\alpha(a) = \left(\frac{a}{1}\right), \qquad \text{for every } a \text{ in } D$$

is an isomorphism.

Proof: We have to show that α is (1) surjective (onto S), (2) injective (1–1), and (3) preserves the operation in D.

1. To show that α is surjective, we choose any element $(a/1)$ in S. This implies that $a \in D$ and by the definition of α,

$$\alpha(a) = \left(\frac{a}{1}\right).$$

Since we have found a preimage for any element in S, α is surjective.

2. To show that α is injective, choose two elements $(a/1) = \alpha(a)$ and $(b/1) = \alpha(b)$ in S. If $(a/1) = (b/1)$, we claim that $a = b$, since $(a/1) = (b/1)$ implies $a/1 = b/1$ or $a \cdot 1 = 1 \cdot b$. Therefore two equal image elements come from two equal elements in the domain and α is injective.

3. We must show that the addition and multiplication in D is preserved. Choose two elements a and b in D. Then

$$\alpha(a + b) = \left(\frac{a+b}{1}\right)$$

$$= \left(\frac{a \cdot 1 + 1 \cdot b}{1 \cdot 1}\right) \quad \text{(since 1 is the identity of } D\text{)}$$

$$= \left(\frac{a}{1}\right) + \left(\frac{b}{1}\right)$$

$$= \alpha(a) + \alpha(b).$$

Also,

$$\alpha(ab) = \left(\frac{ab}{1}\right)$$

$$= \left(\frac{ab}{1 \cdot 1}\right)$$

$$= \left(\frac{a}{1}\right)\left(\frac{b}{1}\right)$$

$$= \alpha(a) \cdot \alpha(b)$$

which completes our proof. ∎

EXERCISES

1. Prove that if n is not prime, then Z_n is not a field.
2. Write out addition and multiplication tables for Z_3, Z_5, Z_7.
3. Prove that any field is
    a. an integral domain
    b. a Euclidean ring.

4. Show that any finite integral domain is a field.
5. Show that the multiplication of equivalence classes, $\left(\dfrac{a}{b}\right) \cdot \left(\dfrac{c}{d}\right) = \left(\dfrac{ac}{bd}\right)$, is well defined.
6. Complete the proof of Theorem 13.3.
7. What is the field of quotients of $\{n + m\sqrt{-1} \mid n, m \in Z\}$, the Gaussian integers?
8. Prove that $S = \left\{ \left(\dfrac{a}{1}\right) \mid a \in D \right\}$ is a subring of

$$F = \left\{ \left(\dfrac{a}{b}\right) \mid a, b \in D, b \neq 0 \right\}.$$

9. If D and D' are two isomorphic integral domains, prove that their field of quotients, F and F', are isomorphic also.

14
Subfields and extension fields

In this chapter we will be investigating properties of subfields and properties of the underlying field that depend on the subfield.

DEFINITION
A subring F' of a field $(F, +, \cdot)$ is called a **subfield** if $(F', +, \cdot)$ is itself a field.

In studying groups and rings, subgroups and subrings played an important part. In the study of fields, subfields will be a key concept.

In Chapter 11 we defined the notion of the characteristic of a ring. We repeat the definition and the major results since we will be using them here.
1. For any ring R, the characteristic of R is the least positive integer n such that $nx = 0$ for all x in R. If no such integer exists, the characteristic of R is zero.
2. The characteristic of R, a ring with identity, is k if and only if k is the least positive integer such that $k \cdot 1 = 0$.
3. The characteristic of an integral domain or a field is either 0 or a prime number.

Let $(F, +, \cdot)$ be a field with multiplicative identity 1. Then the additive order of 1 is either some integer k or infinity. If the additive order of 1 is finite, then this additive order is simply the characteristic of F and must be prime by result 3 above. If the additive order of 1 is infinite, then the characteristic of F is zero. Our first results tell us about an important subfield of any field.

Subfields and extension fields

THEOREM 14.1

Let $(F, +, \cdot)$ be a field with multiplicative identity 1. If the characteristic of F is p, a prime, then the set

$$F_p = \{0, 1, 1+1, 1+1+1, \ldots, (p-1)\cdot 1\}$$

is a subfield of F which is isomorphic to the field Z_p. Furthermore every other subfield of F must contain F_p.

Proof: The set $F_p = \{0, 1\cdot 1, 2\cdot 1, 3\cdot 1, \ldots, (p-1)\cdot 1\}$ where $i\cdot 1$ denotes 1 added to itself i times. Since the characteristic of F is p, addition of elements in F_p must be performed modulo p. (This is left as an exercise.)

If $i\cdot 1, j\cdot 1 \in F_p$, we use the distributive laws to compute

$$(i\cdot 1)\cdot(j\cdot 1) = \underbrace{(1+1+\cdots+1)}_{i \text{ times}} \cdot \underbrace{(1+1+\cdots+1)}_{j \text{ times}}$$

$$= \underbrace{(1+1+\cdots+1)}_{ij \text{ times}}.$$

By the division algorithm, there exist integers k and r such that $ij = pk + r$ where $0 \le r < p$. Therefore we have

$$(i\cdot 1)(j\cdot 1) = ij\cdot 1$$
$$= kp\cdot 1 + r\cdot 1$$
$$= 0 + r\cdot 1$$
$$= r\cdot 1$$

since the characteristic of the field is p. Therefore the multiplication in F_p is multiplication modulo p.

The function $\alpha: Z_p \to F_p$ defined by $\alpha(i) = i\cdot 1$ for every i in Z_p is an isomorphism.

Finally, if K is a subfield of F, then the multiplicative identity $1 \in K$ (Exercise 1 of this chapter) and all the sums $1, 1+1, 1+1+1, \ldots, (p-1)\cdot 1$ must be in K also so that $F_p \subseteq K$. ∎

Note that F_p can be thought of as the subfield generated by 1; that is, $F_p = \langle 1 \rangle$.

THEOREM 14.2

If the characteristic of a field F is zero, then the set

$$F_\infty = \{(n\cdot 1)(m\cdot 1)^{-1} \mid n, m \in Z, m \ne 0\}$$

is a subfield of F which is isomorphic to the field of rational numbers. Every other subfield of F must contain F_∞.

Proof: If the characteristic of a field F is zero, then any subfield containing the multiplicative identity 1 must contain $\underbrace{1 + 1 + \cdots + 1}_{n \text{ times}} = n \cdot 1$, $(-n \cdot 1)$, and $(n \cdot 1)^{-1}$ for all positive integers n. Also, it must contain all products $(n \cdot 1)(m \cdot 1)^{-1}$ for $m \neq 0$. The theorem asserts that this set F_∞ forms a subfield isomorphic with the rationals. We leave as an exercise that the mapping α from the rationals to F_∞ given by

$$\alpha\left(\frac{n}{m}\right) = (n \cdot 1)(m \cdot 1)^{-1}$$

is an isomorphism and that F_∞ is a field.

If K is a subfield of F, then F_∞ is a subfield of K for the reasons stated in the proof of Theorem 14.2. ∎

Thus these theorems say that any field F, depending on its characteristic, has either Z_p or the rationals as a subfield, and that the minimal subfield of F is either Z_p or the rationals.

DEFINITION
The subfield of F given by either Theorem 14.1 or 14.2 is called the **prime field** of F.

Having found that a familiar subfield is contained in any field, let us investigate the original field more closely.

DEFINITION
Let E be a field and F a subfield of E. Then E is called an **extension field** of F.

EXAMPLES 14.1
1. The field of real numbers is an extension field of the rationals.
2. Any field is an extension field of itself.
Consider the polynomial

$$p(x) = a_0 + a_1 x + a_2 x^2 + \cdots + a_n x^n$$

with coefficients $a_0, a_1, \ldots, a_n$ in a field F. For any symbol t, we define

$$p(t) = a_0 + a_1 t + a_2 t^2 + \cdots + a_n t^n.$$

If t is in F or an extension field of F, the expression $p(t)$ has a value. Otherwise t will be taken as indeterminate.

Subfields and extension fields

DEFINITION

If F is a field and E is an extension field of F and $t \in E$, then
1. t is called a **root** of the polynomial $p(x)$ if $p(t) = 0$. t is said to satisfy p.
2. If t satisfies a polynomial with coefficients in F, we call t **algebraic** over F.
3. If there is no polynomial over F which is satisfied by t, we call t **transcendental** over F.

EXAMPLE 14.2

If $\mathbb{R}$ is the field of real numbers and Q is the field of rational numbers, then $\sqrt{2}$ is algebraic over Q since $\sqrt{2}$ satisfies the polynomial $p(x) = x^2 - 2$. On the other hand, π is transcendental over F since there is no polynomial with rational coefficients for which π is a root. This is a famous result which we will not prove in this book.

Now let E be an extension of F and let t be algebraic over F. Then t satisfies a polynomial $p(x)$ over F of minimal degree.

DEFINITION

If t is algebraic over F, then a polynomial p of minimal degree for which $p(t) = 0$ is called a **minimal polynomial** satisfied by t.

LEMMA

A minimal polynomial $p(x)$ of an algebraic element t is irreducible over F.

Proof: We prove this by contradiction. If

$$p(x) = p_1(x) \cdot p_2(x)$$

where neither p_1 nor p_2 are units, then both p_1 and p_2 have degree less than that of p. We see that

$$p(t) = p_1(t)p_2(t) = 0$$

which implies that $p_1(t) = 0$ or $p_2(t) = 0$ and this contradicts the hypothesis that p was a minimal polynomial of t. ∎

THEOREM 14.3

Let E be an extension of F and let $t \in E$ be algebraic over F. Then the set of all polynomials in $F[x]$ which are satisfied by t is exactly the ideal $(p(x))$ where $p(x)$ is a minimal polynomial satisfied by t.

Proof: We will show that (1) every polynomial in $(p(x))$ is satisfied by t and (2) every polynomial satisfied by t is in $(p(x))$.

1. Let $q(x)$ be any polynomial in $(p(x))$. Then $q(x) = p(x) \cdot r(x)$. $p(t) = 0$ implies $q(t) = 0$ so that t also satisfies $q(x)$.

2. Let $p_1(x)$ be any polynomial satisfied by t, $p_1(t) = 0$. Then, by the division algorithm for polynomials (Theorem 12.10) we know there exist polynomials $q(x)$ and $r(x)$ such that

$$p_1(x) = q(x)p(x) + r(x) \quad \text{where } r(x) = 0$$

or

$0 \le \text{degree } r(x) < \text{degree } p(x)$.

Evaluating for $x = t$, we get

$$p_1(t) = q(t)p(t) + r(t)$$
$$0 = q(t) \cdot 0 + r(t)$$
$$0 = r(t)$$

and degree $r(x)$ < degree $p(x)$. This is impossible unless $r(x) = 0$ because the degree of $p(x)$ was assumed to be minimal. Therefore we see that $p_1(x) = q(x) \cdot p(x) \in (p(x))$. ∎

COROLLARY

$F[x]/(p(x))$ is a field where $p(x)$ is a minimal polynomial of an algebraic element t over the field F.

Proof: $p(x)$ is irreducible by the lemma and therefore $F[x]/(p(x))$ is a field by Theorem 12.11. ∎

We are considering an extension E of a field F and an element t which is algebraic over F with minimal polynomial $p(x)$ over F. There are subfields of E which contain both F and t. E itself is an example of one.

DEFINITION

Let $F(t)$ be the intersection of all those subfields of E which contain F and t.

Then $F \subseteq F(t)$, $t \in F(t)$, and $F(t)$ is a subfield of E. (Why?) In fact, $F(t)$ is the smallest subfield of E containing both F and t.

DEFINITION
Let

$\overline{F(t)} = \{a_0 + a_1 t + \cdots + a_n t^n \mid a_i \in F, i = 0, 1, \ldots, n$ and

n is any nonnegative integer$\}$.

Subfields and extension fields

Our next result ties these ideas together and gives a very useful handle on the subfields of E by identifying some of them with the fields $F[x]/(p(x))$ where $p(x)$ is an irreducible polynomial.

THEOREM 14.4
Let E be an extension of F, and let t be an algebraic element over F satisfying the minimal polynomial $p(x)$ over F. Let $F(t)$ and $\overline{F(t)}$ be as defined above. Then
1. $F(t) = \overline{F(t)}$ and
2. $F(t)$ is isomorphic with $F[x]/(p(x))$.

Proof: Any subfield containing F and t will contain all sums of the form $a_0 + a_1 t + a_2 t^2 + \cdots + a_n t^n$ so it is immediate that $\overline{F(t)} \subseteq F(t)$.

To show that $F(t) \subseteq \overline{F(t)}$ and that $F(t)$ is isomorphic with $F[x]/(p(x))$, we shall show that $F[x]/(p(x))$ and $\overline{F(t)}$ are isomorphic as rings. We shall then know that $\overline{F(t)}$ is a subfield of $F(t)$ containing F and t. By the minimality of $F(t)$, it must be true that $F(t) = \overline{F(t)}$.

To construct our isomorphism, consider the mapping $\alpha: F[x] \to \overline{F(t)}$ defined by

$$\alpha(a_0 + a_1 x + \cdots + a_n x^n) = a_0 + a_1 t + \cdots + a_n t^n.$$

Clearly this is a surjective mapping from $F[x]$ onto $\overline{F(t)}$. To verify that α is a ring homomorphism, show that

$$\alpha[(a_0 + a_1 x + \cdots + a_n x^n) + (b_0 + b_1 t + \cdots + b_m t^m)]$$
$$= a_0 + a_1 t + \cdots + a_n t^n + b_0 + b_1 t + \cdots + b_m t^m$$

and

$$\alpha[(a_0 + a_1 x + \cdots + a_n x^n) \cdot (b_0 + b_1 x + \cdots + b_m x^m)]$$
$$= (a_0 + a_1 t + \cdots + a_n t^n) \cdot (b_0 + b_1 t + \cdots + b_m x^m).$$

The computations are straightforward and are left as an exercise.

What is the kernel of α? If $q(x) = a_0 + a_1 x + \cdots + a_n x^n$,

$$\alpha(q(x)) = \alpha(a_0 + a_1 x + \cdots + a_n x^n)$$
$$= 0$$

if and only if $a_0 + a_1 t + \cdots + a_n t^n = q(t) = 0$. By Theorem 14.3 this holds if and only if $q(x) \in (p(x))$. Therefore $\ker(\alpha) = (p(x))$. Finally, by Theorem 11.10, we conclude that $F[x]/(p(x))$ and $\overline{F(t)}$ are isomorphic, which completes the proof of the theorem. ∎

EXERCISES

1. Let F be a field and let F' be a subfield. Show that F' has the same additive and multiplicative identities as F.
2. a. If p is any prime integer, show that Z_p has no proper subfields.
   b. Show that the rationals have no proper subfields.
3. Complete the proofs of Theorem 14.1 and Theorem 14.2.
4. Let F be a field with a finite number of elements. Show that the order of F must be a power of its characteristic.
5. Show that the intersection of two fields is always a field.
6. Determine whether the following real numbers are algebraic or transcendental over the field of rationals:
   a. $\sqrt{2} + 2$
   b. $\sqrt{2} - \sqrt{3}$
   c. $\sqrt{a} + \sqrt[3]{b}$, where a and b are rational
   d. $\pi - 1$
   e. $\frac{1}{2}$.
7. Show that the product of two algebraic elements is algebraic. Can you say anything about the sum?
8. Classify all minimal polynomials of $\sqrt{2}$ over the rationals. Do the same for $\sqrt[3]{4}$ over the rationals.
9. Prove $\sqrt{2}$ is not rational. (*Hint*: Prove this by contradiction. Assume $\sqrt{2} = n/m$ where $(n, m) = 1$.)
10. Prove that $\sqrt{n}$ is not rational if n is a positive integer that is not a perfect square.
11. Suppose t is algebraic over F and satisfies a minimal polynomial of degree n. Show that t satisfies a unique irreducible polynomial of degree n whose nth coefficient is 1.
12. Verify that the mapping α defined in the proof of Theorem 14.4 is a homomorphism.
13. a. If Q is the field of rationals
       (i) find $Q(\sqrt{2})$.
       (ii) find $Q(\sqrt{3})$. Simplify your answers as much as possible.
    b. Find $F(\sqrt{2})$ and $F(\sqrt{3})$ for any field F.
    c. Describe $(F(\sqrt{2}))(\sqrt{3})$.
    d. Show $(F(t_1))(t_2) = (F(t_2))(t_1)$ where t_1 and t_2 are any two algebraic elements over F.
14. Let F be a subfield of E and let t_1 and t_2 be algebraic over F. Suppose that $p(x)$ is a minimal polynomial of t_2 over F and $q(x)$ is the minimal polynomial of t_2 over $F(t_1)$. Show that $q(x)$ divides $p(x)$ in $F(t_1)[x]$.
15. a. If $p(x)$ is an irreducible polynomial, $p(x) = a_0 + a_1 x + \cdots + a_n x^n$, over F, show that $F[x]/(p(x))$ can be considered as the set of formal expressions of the form
    $$q(t) = b_0 + b_1 t + \cdots + b_{n-1} t^{n-1}$$
    where $b_i \in F$, $i = 0, 1, \ldots, n-1$ and where addition and multiplication are defined $\mod(p(t))$.
    b. Use part a to describe $F(\sqrt{2}) = F[x]/(x^2 - 2)$.

15

Roots of polynomials in F[x]

We now have a sense of what happens if we have an element t of E which is algebraic over F. That is, the field $F(t)$ is identified with $F[x]/(p(x))$, where $p(x)$ is the minimal polynomial of t over F. Next, let us turn the situation around and start with a field F and an irreducible polynomial $p(x)$ in $F[x]$. We ask if we can find extensions of F which contain roots of $p(x)$. It turns out that the field $E = F[x]/(p(x))$ has this property.

THEOREM 15.1
Let F be a field and let $p(x)$ be any irreducible polynomial in $F[x]$. The field $E = F[x]/(p(x))$ has the following properties:
1. E is an extension of F in the sense that E contains a subfield isomorphic to F.
2. E contains an element that satisfies the polynomial $p(x)$ over F.

Proof: 1. Consider the subset

$$\bar{F} = \{(p(x)) + a \mid a \in F\}$$

of E. $\bar{F}$ is a subfield of E which is isomorphic to F under the mapping $a \to (p(x)) + a$. In this sense E is an extension of F (that is, of $\bar{F}$). The details are left as an exercise.

2. Consider a polynomial $p(x)$ over F,

$$p(x) = a_0 + a_1 x + \cdots + a_n x^n$$

where $a_i \in F$ for $i = 0, 1, \ldots, n$. This polynomial corresponds to the polynomial $\bar{p}(x)$ over $\bar{F}$,

$$\bar{p}(x) = \bar{a}_0 + \bar{a}_1 x + \cdots + \bar{a}_n x^n$$

where $\bar{a}_i = (p(x)) + a_i$ for $i = 0, 1, \ldots, n$.

Let $\bar{x} = (p(x)) + x \in E$. We claim that $\bar{x}$ satisfies $\bar{p}(x)$ over $\bar{F}$. Substituting $\bar{x}$ for x, we get

$$\bar{p}(\bar{x}) = \bar{a}_0 + \bar{a}_1 \bar{x} + \cdots + \bar{a}_n \bar{x}^n$$
$$\in (p(x)) + p(x)$$
$$= 0 \quad \text{in } E.$$

Therefore $p(x)$ (or $\bar{p}(x)$) has a root in E. ∎

EXAMPLE 15.1
Let Q be the field of rationals. The polynomial $p(x) = x^2 - 2$ is irreducible over Q. By the theorem, the field $E = Q[x]/(x^2 - 2)$ is an extension of F which contains a root of $x^2 - 2$. By Theorem 14.4, $Q[x]/(x^2 - 2) = Q(\sqrt{2}) = \{a + b\sqrt{2} \mid a, b \in Q\}$. Therefore $Q(\sqrt{2})$ is an extension field of Q which contains a root of the irreducible polynomial $x^2 - 2 \in Q[x]$.

Using this theorem repeatedly, we can find an extension of F which has all roots of $p(x)$. Before we do that, we have to take a closer look at certain aspects of the polynomial ring $F[x]$. We need to know some of the properties of the roots of a polynomial.

THEOREM 15.2
Let F be a field, let E be an extension of F, and let $p(x)$ be a polynomial over F. Then an element $a \in E$ satisfies $p(x)$ if and only if $(x - a)$ divides $p(x)$ in $E[x]$.

Proof: 1. If $(x - a)$ divides $p(x)$, then there is a $q(x)$ in $E[x]$ with

$$p(x) = (x - a)q(x).$$

By substitution

$$p(a) = (a - a) \cdot q(a)$$
$$= 0 \cdot q(a)$$
$$= 0.$$

2. Next assume that $p(a) = 0$. By Theorem 12.10 we have

$$p(x) = q(x) \cdot (x - a) + r(x)$$

where $r(x) = 0$ or degree $r(x) <$ degree $(x - a) = 1$. Therefore $r(x) = 0$ or $r(x) = c$, a constant. Again, substituting $x = a$, we see that

$$p(a) = q(a) \cdot (a - a) + c$$

or

$0 = q(a) \cdot 0 + c$

$0 = c$.

Therefore $p(x) = (x - a) \cdot q(x)$ as the theorem asserts. ∎

COROLLARY
A polynomial of degree n over a field can have at most n roots.

The proof is left as an exercise.

DEFINITION
If F is a field, E is an extension of F, and $p(x)$ is a polynomial over F, then for a positive integer n, an element $a \in E$ is said to be a **root of multiplicity n** if

$$p(x) = (x - a)^n q(x)$$

for some polynomial $q(x)$ over E.

If $n > 1$, then a is called a **multiple root** or a **repeated root** of $p(x)$.

THEOREM 15.3
Let $p(x) = a_0 + a_1 x + \cdots + a_n x^n$ be a polynomial over a field F. If we define the formal derivative

$$p'(x) = a_1 + 2a_2 x + \cdots + n a_n x^{n-1}$$

as in the calculus, then an element a is a repeated root of $p(x)$ if and only if a is also a root of $p'(x)$.

Proof: 1. If a is a repeated root of $p(x)$, then $p(x) = (x - a)^n q(x)$ for some $n > 1$ and

$$p'(x) = n(x - a)^{n-1} q(x) + (x - a)^n q'(x).$$

We see that $p'(a) = 0$.

2. If a is a single root of $p(x)$, then $p(x) = (x - a)q(x)$ where $(x - a)$ does not divide $q(x)$. Then

$$p'(x) = q(x) + (x - a)q'(x).$$

and $(x - a)$ does not divide $p'(x)$. ∎

Using Theorems 15.1 and 15.2 we can prove the existence of an extension field which contains all the roots of a polynomial over F.

THEOREM 15.4
Let F be a field and let $p(x)$ be a polynomial (not necessarily irreducible) in $F[x]$. Then there exists an extension field E of F in which $p(x)$ factors completely into (linear) factors of degree 1.

Proof: The proof can be done by induction on the degree of $p(x)$. Suppose the degree of $p(x)$ is n.

If $n = 1$, there is nothing to prove since $p(x) = ax + b$.

Assume that the theorem is true for all polynomials of degree $\leq k - 1$, and assume that $p(x)$ is a polynomial of degree k.

Case 1: If $p(x)$ is irreducible, then Theorem 15.1 states that there is an extension field E_1 of F which contains a root a of $p(x)$ and we can write

$$p(x) = (x - a) \cdot q(x) \text{ in } E_1.$$

Case 2: If $p(x)$ is reducible, write $p(x)$ as a product of its irreducible factors in $F[x]$,

$$p(x) = p_1(x)p_2(x) \cdots p_l(x)$$

where each $p_i(x)$, $i = 1, 2, \ldots, l$, has degree less than k. Then Theorem 15.1 says that there is an extension field E_1 of F such that E_1 contains a root a of $p_1(x)$. By Theorem 15.2, $p_1(x) = (x - a_1)q_1(x)$ where $q_1(x) \in E_1[x]$. We can write

$$p(x) = (x - a)q_1(x)p_2(x) \cdots p_l(x)$$
$$= (x - a)q(x).$$

In either case, we can write $p(x) = (x - a)q(x)$ where

degree $q(x)$ = degree $p(x) - 1$
$\qquad\qquad = k - 1$.

Our induction hypothesis says that we can find an extension field E of E_1 such that $q(x)$ can be factored into linear factors in $E[x]$. Since $F \subseteq E_1 \subseteq E$, $p(x) = (x - a)q(x)$ can also be factored into linear factors in E. ∎

EXAMPLE 15.2
Let Q = rational numbers, $p(x) = (x^2 - 2)(x^2 - 3)$. Then over the field $\mathbb{R}$, the real numbers, $p(x)$ factors as

$$p(x) = (x - \sqrt{2})(x + \sqrt{2})(x - \sqrt{3})(x + \sqrt{3}).$$

Over the field $Q(\sqrt{2})$

$$p(x) = (x - \sqrt{2})(x + \sqrt{2})(x^2 - 3).$$

Over the field $Q(\sqrt{2}, \sqrt{3}) = (Q(\sqrt{2}))(\sqrt{3})$, $p(x)$ factors as over $\mathbb{R}$. So both $\mathbb{R}$ and $Q(\sqrt{2}, \sqrt{3})$ are extension fields of the field Q where the polynomial $(x^2 - 2)(x^2 - 3)$ factors completely, but $Q(\sqrt{2})$ does not work.

The theorem states that there is always at least one extension field E such that a polynomial has all linear factors in $E[x]$. Our next step is to see that there is always a minimal extension in which $p(x)$ can be completely factored.

DEFINITION

If F is a field, $p(x)$ a polynomial over F, and E an extension of F such that $p(x)$ factors completely into linear factors in $E[x]$ and such that $p(x)$ does not factor completely into linear factors in any proper subfield of E, then E is called a **splitting field** for $p(x)$ over F.

THEOREM 15.5

Let F be a field and let $p(x)$ be a polynomial over F. Then there exists a splitting field for $p(x)$ over F.

Proof: Let E be an extension in which $p(x)$ factors completely and let $\bar{E}$ be the intersection of all the subfields of E in which $p(x)$ factors completely. Then $\bar{E}$ is a splitting field of $p(x)$ over F. ∎

We are now in a position to use our results on splitting fields to completely determine all finite fields.

THEOREM 15.6

Let F be a finite field. Then F has exactly p^n elements in it for some prime p and some positive integer n.

Proof: Let F be a finite field and let F have characteristic p for some prime p. Then for any $x \in F$, $px = 0$. That is, every nonzero element of F has additive order p.

By contradiction, assume the order of F is not p^n; then some other prime q divides the order of F. By Cauchy's Theorem for Abelian Groups, there is an element of additive order q. This is not possible. Therefore the order of F must be p^n. ∎

We will use the material developed thus far to classify finite fields. We need a computational lemma first.

LEMMA

Let F be a field of characteristic p. Then for any $a, b \in F$ and positive integer n,
1. $(a+b)^{p^n} = a^{p^n} + b^{p^n}$ and
2. if $a \neq 0$, $(a^{-1})^{p^n} = (a^{p^n})^{-1}$.

Proof: We will prove part 1 by induction on n. For $n=1$,

$$(a+b)^p = a^p + pa^{p-1}b + \frac{p(p-1)}{1\cdot 2} a^{p-2}b^2 + \cdots + pab^{p-1} + b^p$$

by the Binomial Theorem. Every term except the first and the last has p as a coefficient and therefore each term is zero since p is the characteristic of F. We are left with

$$(a+b)^p = a^p + b^p$$

and the lemma is true for $n=1$.

Assume that $(a+b)^{p^n} = a^{p^n} + b^{p^n}$ for all $n \leq k$. Consider

$$\begin{aligned}(a+b)^{p^{k+1}} &= ((a+b)^{p^k})^p \\ &= (a^{p^k} + b^{p^k})^p \\ &= (a^{p^k})^p + (b^{p^k})^p \\ &= a^{p^{k+1}} + b^{p^{k+1}}.\end{aligned}$$

Since the induction hypothesis implies that the statement is true for $n = k+1$, it is true for all positive integers n. The second statement follows from the Laws of Exponents. ∎

We now prove our major result.

THEOREM 15.7

For any prime p and any positive integer n, there is a finite field with exactly p^n elements.

Proof: Let $F = Z_p$, the field with p elements, and let E be a splitting field of the polynomial $q(x) = x^{p^n} - x$ over F. We wish to show that E has exactly p^n elements in it. First we will prove that the roots of $q(x)$ form a field. This implies that E consists precisely of all the roots of $q(x)$. (Why?)

Note that a is a root of $q(x) = x^{p^n} - x$ if and only if $a^{p^n} - a = 0$ or $a^{p^n} = a$.

1. If a and b are roots of $q(x)$, then

$$\begin{aligned}(a-b)^{p^n} &= a^{p^n} - b^{p^n} \quad \text{(using the lemma)} \\ &= a - b \quad \text{(by the property of a root of } q(x)\text{)}.\end{aligned}$$

Therefore $(a - b)$ is also a root of $q(x)$ and the roots of $q(x)$ form an additive group.

2. If a and b are roots of $q(x)$, then

$$(ab)^{p^n} = a^{p^n} b^{p^n}$$
$$= ab$$

and ab is also a root.

3. If $a \neq 0$ is a root of $q(x)$, then

$$(a^{-1})^{p^n} = (a^{p^n})^{-1} \quad \text{(by the lemma)}$$
$$= a^{-1}$$

and a^{-1} is also a root.

Thus the roots of $q(x)$ form a field, proving that E is precisely the roots of $q(x)$. Now, by the corollary to Theorem 15.2, $q(x)$ can have at most p^n roots so the order of E is at most p^n.

To see that the order of E is exactly p^n we need to show that $q(x)$ can have no repeated roots. We use Theorem 15.3.

$$q'(x) = p^n x^{p^n - 1} - 1$$
$$= -1$$

since Z_p has characteristic p and $p^n x^{p^n} = 0$. Since $q'(x) = -1$ has no roots, $q(x)$ has no repeated roots and E has exactly p^n elements. ∎

This result can be extended. It is not too difficult to prove that any two splitting fields for a polynomial over a field are isomorphic and it follows that any two fields of order p^n are isomorphic. This leads to the usual notation GF(p^n) for the unique field of order p^n. (The GF stands for Galois Field after the mathematician Galois who first studied them.) With this last result, all finite fields have been completely classified.

The proof that any two splitting fields for a polynomial $p(x)$ over a field F are isomorphic is not difficult; however it will not be given here. There are many good treatments of the subject available where you can find a proof of this result and related results on the isomorphism of certain types of extension fields.

EXERCISES

1. Let Q be the rational numbers, $p(x) = x^2 - 2$. Construct the field $Q[x]/(p(x))$ and compare it with the subfield $Q(\sqrt{2})$ of the real numbers. Show that the two fields are isomorphic. What is the element of $Q[x]/(p(x))$ which satisfies $p(x)$?

2. Prove that a polynomial of degree n can have at most n roots in any extension field.
3. In the proof of Theorem 15.1,
   a. Show that $\bar{F}$ is a field.
   b. Show that $\alpha: F \to \bar{F}$ defined by $\alpha(a) = (p(x)) + a$ is an isomorphism.
4. Write out a detailed proof of Theorem 15.5.
5. Find splitting fields for the following polynomials over the rational numbers.
   a. $x^2 - 2$
   b. $(x^2 - 2)(x^2 - 3)$
   c. $x^2 - 4$
   d. $x^4 - 1$.
   e. $x^3 - 1$
6. Let $F = Z_p$, the finite field containing p elements. Find an irreducible polynomial of degree 2 over Z_p and use this to construct a field of p^2 elements. Do this explicitly for the cases $p = 2, 3$.
7. Show that every field F of order p^n has the property that for every $x \in F$, $x^{p^n} = x$.
8. Show that any field of order p^n is a splitting field of $q(x) = x^{p^n} - x$ over a subfield isomorphic to Z_p.
9. Assume that any two splitting fields of a polynomial over a field are isomorphic and show that any two fields of order p^n are isomorphic.
10. Show that the multiplicative group of $GF(p^n)$ is cyclic.
11. Defining the formal derivative as in Theorem 15.3, prove that for any polynomials $p(x)$ and $q(x)$, it is true that
    a. $(p(x) + q(x))' = p'(x) + q'(x)$ and
    b. $(p(x)q(x))' = p'(x)q(x) + p(x)q'(x)$.

16
Galois theory

Our last topic in this book will be a brief discussion of some of the elements of Galois Theory which leads to the question of solving polynomials by extracting roots.

For example, the general quadratic $ax^2 + bx + c$ is known to have roots $(-b \pm \sqrt{b^2 - 4ac})/2a$. Similarly, the general polynomials of degrees 3 and 4 can be solved by taking square roots, cube roots (radicals), and so on. There are formulas which give the solutions of these polynomials in terms of radicals.

It turns out that for polynomials of degree 5 or more, no such general formulas exist. Thus it is impossible to find all solutions to a general polynomial of degree greater than 4 by adding, subtracting, multiplying, and dividing various combinations of square, cube, fourth, fifth, etc. roots of the coefficients of the polynomial.

It requires a great deal of technique to prove the "unsolvability" by radicals of the general polynomial of degree 5 or greater. In this chapter we sketch some of the proof, introduce the new techniques required, and try to make clear the complexity of the situation.

Galois Theory has also been used to solve some famous geometric problems. An application of this theory shows that it is impossible to trisect an angle or to square a circle by ruler and compass construction.

Before continuing we would like to warn the reader that the remainder of this chapter is quite dense with new material and contains a number of results without proof. The proofs require more sophistication than the previous material, but we feel it worthwhile to sketch some of the major results so that you could get a feel for one of the more beautiful applications of Field Theory and Group Theory. A careful reading of the next few pages should be quite rewarding even though some of the key proofs have been omitted.

DEFINITION

A polynomial $p(x)$ over a field F is **solvable by radicals** if there is a "tower" of fields,

$$F_0 \subseteq F_1 \subseteq F_2 \subseteq \cdots \subseteq F_k$$

where
1. all the roots of $p(x)$ are in F_k, and
2. for each i, $1 \le i \le k$, there is a positive integer n_i and an element $r_i^{n_i} \in F_{i-1}$ such that $F_i = F_{i-1}(r_i)$.

That is, by adjoining to F the n_1st root of some element in F we get a field F_1. Then adjoining to F_1 the n_2nd root of some element in F_1 we get a field F_2. Continuing, we end up with a field F_k which contains all the roots of $p(x)$ and hence contains a splitting field of $p(x)$.

EXAMPLE 16.1

Let $p(x) = x^2 + 2x + 5$ be a polynomial over the real numbers. The roots of $p(x)$ are $(-2 \pm \sqrt{-16})/2 = -1 \pm 2\sqrt{-1}$. Thus $F_1 = \mathbb{R}(\sqrt{-1})$ contains both roots of $p(x)$ and $(\sqrt{-1})^2 = -1 \in \mathbb{R}$.

What we shall do is associate with a polynomial $p(x)$ over a field F, a group called the **Galois group.** The Galois group is very important in many aspects of field theory. Its importance in the area of solvability by radicals is that $p(x)$ is solvable by radicals if and only if its Galois group is a solvable group. The final step in our argument is to show that there always exist polynomials of degree n whose Galois groups are S_n. Since S_n is not solvable when $n \ge 5$, the chain of logical inference will be complete.

We are not going to do the whole thing, but enough to get an idea of the development, and a good indication of the structure of the Galois group. The first notion we discuss is the concept of the degree of an extension K over a subfield F. This is the ordinary dimension if you have studied vector spaces. We will develop the subject from the beginning.

DEFINITION

Let K be an extension of F and let $x_1, x_2, \ldots, x_n$ be elements of K. We say that the set $\{x_1, x_2, \ldots, x_n\}$ is **linearly dependent** over F if there are elements $a_1, a_2, \ldots, a_n$ in F, not all zero, such that $a_1 x_1 + a_2 x_2 + \cdots + a_n x_n = 0$. If no elements a_i, $i = 1, 2, \ldots, n$, can be found, then the set $\{x_1, x_2, \ldots, x_n\}$ is called **linearly independent.**

EXAMPLE 16.2

If $K = Q(\sqrt{2})$, where Q is the field of rational numbers, let $x_1 = 1 + \sqrt{2}$, $x_2 = 1 - \sqrt{2}$, and $x_3 = 2\sqrt{2}$. For the rational numbers $a_1 = 1$, $a_2 = -1$, $a_3 = -1$,

$$a_1 x_1 + a_2 x_2 + a_3 x_3 = 1(1 + \sqrt{2}) + (-1)(1 - \sqrt{2}) + (-1)(2\sqrt{2})$$
$$= 0.$$

Therefore $\{x_1, x_2, x_3\}$ is linearly dependent.

On the other hand, the set $\{x_1, x_2\}$ is linearly independent for if a_1, $a_2 \in Q$ and

$$a_1(1 + \sqrt{2}) + a_2(1 - \sqrt{2}) = 0$$

then $a_1 = a_2 = 0$. Similarly the sets $\{x_1, x_3\}$ and $\{x_2, x_3\}$ are linearly independent.

We can generalize these results. For $K = Q(\sqrt{2})$, as above, we can show that any three elements of K are linearly dependent over Q. Let

$$x_1 = r_1 + r_2\sqrt{2}$$
$$x_2 = s_1 + s_2\sqrt{2}$$
$$x_3 = t_1 + t_2\sqrt{2}$$

where $r_1, r_2, s_1, s_2, t_1, t_2 \in Q$.

We want to find elements a_1, a_2, a_3 in Q such that

$$a_1 x_1 + a_2 x_2 + a_3 x_3 = 0.$$

The elements a_1, a_2, a_3 must satisfy the following:

$$a_1(r_1 + r_2\sqrt{2}) + a_2(s_1 + s_2\sqrt{2}) + a_3(t_1 + t_2\sqrt{2})$$
$$= a_1 r_1 + a_2 s_1 + a_3 t_1 + (a_1 r_2 + a_2 s_2 + a_3 t_2)\sqrt{2}$$
$$= 0.$$

This implies that

$$a_1 r_1 + a_2 s_1 + a_3 t_1 = 0$$

and

$$a_1 r_2 + a_2 s_2 + a_3 t_2 = 0.$$

This is a homogeneous system of two equations in the three unknowns a_1, a_2, a_3. Therefore there is a solution for a_1, a_2, and a_3 *other* than the trivial solution $a_1 = 0, a_2 = 0, a_3 = 0$, and we see that the set $\{r_1 + r_2\sqrt{2}, s_1 + s_2\sqrt{2}, t_1 + t_2\sqrt{2}\}$ is linearly dependent.

We give this concept a name.

DEFINITION

If K is an extension of F, the **degree of K over F,** written $[K : F]$, is the smallest integer n with the property that any set of $n + 1$ elements of K are linearly dependent.

That is, there exists a set of n linearly independent elements of K over F, while there does not exist a set of $n + 1$ linearly independent elements of K over F.

EXAMPLE 16.3

$[Q(\sqrt{2}) : Q] = 2$.

DEFINITION

If, for every integer n, there is a set of n linearly independent elements of K over F, we say that $[K : F] = \infty$.

EXERCISES

1. In Example 16.1 show that the sets $\{x_1, x_2\}$, $\{x_1, x_3\}$, and $\{x_2, x_3\}$ are linearly independent.
2. Give an example of two elements x_1, x_2 of $Q(\sqrt{2})$ which are linearly dependent over Q.
3. If Q is the field of rationals, let $K = Q(\sqrt{r})$ where $r \in Q$ but $\sqrt{r} \notin Q$. Show that $[K : Q] = 2$.
4. a. In $Q(\sqrt[3]{2})$ determine which of the following sets are linearly independent.
    (i) $\{1 - \sqrt[3]{2}, 1 + \sqrt[3]{2^2}, 1\}$
    (ii) $\{1 - \sqrt[3]{2} + \sqrt[3]{4}, 1 + \sqrt[3]{2} - \sqrt[3]{4}\}$
    (iii) $\{1, \sqrt[3]{2}, \sqrt[3]{4}\}$
    (iv) $\{1, 2, 3\}$
    (v) $\{a_1 + b_1\sqrt[3]{2} + c_1\sqrt[3]{4}, a_2 + b_2\sqrt[3]{2} + c_2\sqrt[3]{4},$
    $a_3 + b_3\sqrt[3]{2} + c_3\sqrt[3]{4}, a_4 + b_4\sqrt[3]{2} + c_4\sqrt[3]{4}\}$
    (vi) $\{2 - \sqrt[3]{2} + 3\sqrt[3]{4}, -4 + 2\sqrt[3]{2} - 6\sqrt[3]{4}\}$.
   b. Find $[Q(\sqrt[3]{2}) : Q]$.
5. Let Q be the rational numbers and $p(x) = x^2 - 2$. If $K = Q[x]/(p(x))$, find $[K : Q]$.
6. If F is any field and $p(x)$ is an irreducible polynomial of degree n over F and if $K = F[x]/(p(x))$, show that $[K : F] = n$.
7. What is $[\mathbb{R} : Q]$?
8. Let $[K : F] = n$ and let $\{x_1, \ldots, x_n\}$ be a set of n linearly independent elements over F. Show that *any* element $x \in K$ can be expressed in the form

$$x = a_1 x_1 + a_2 x_2 + \cdots + a_n x_n \quad \text{for some } a_1, a_2, \ldots, a_n \in F.$$

9. Let K be an extension of F and let $\{x_1, \ldots, x_n\}$ be a set of linearly independent elements. Show that if $a_1x_1 + a_2x_2 + \cdots + a_nx_n = b_1x_1 + b_2x_2 + \cdots + b_nx_n$ for $a_i, b_i \in F$, then $a_1 = b_1, a_2 = b_2, \ldots, a_n = b_n$.

10. a. Let K be an extension of F. If x_1, x_2 are linearly independent over F and if

$$y_1 = a_{11}x_1 + a_{12}x_2$$
$$y_2 = a_{21}x_1 + a_{22}x_2$$
$$y_3 = a_{31}x_1 + a_{32}x_2$$

for $a_{ij} \in F$, show that $\{y_1, y_2, y_3\}$ are linearly dependent over F.
b. Generalize this result if you can.

11. Use Exercise 10 to prove the following: Let K be an extension of F and let $x_1, x_2 \in K$ be such that any $x \in K$ can be written in the form

$$x = a_1x_1 + a_2x_2 \quad \text{for some } a_1, a_2 \in F.$$

Show that $[K:F] = 2$.

12. Let $p(x)$ be of degree n over F and let K be the splitting field of $p(x)$ over F. Show that $[K:F] \leq n!$.

Our next step is to look at field automorphisms.

DEFINITION
If K is a field, then α is said to be an **automorphism** of K if α is a bijective (1–1 and onto) map of K onto itself such that
1. $\alpha(x + y) = \alpha(x) + \alpha(y)$
and
2. $\alpha(x \cdot y) = \alpha(x) \cdot \alpha(y)$
for all $x, y \in K$.

The set of all automorphisms of K is denoted by $\mathscr{A}(K)$.

THEOREM 16.1
$\mathscr{A}(K)$ forms a group under composition.
 Proof: See the proof of Theorem 7.4 in Part I. ∎

DEFINITION
If $\alpha \in \mathscr{A}(K)$ and $a \in K$, then a is said to be **fixed** by α if $\alpha(a) = a$.

EXAMPLE 16.4
The identity automorphism fixes every element of K. The identity element is fixed by every automorphism.

THEOREM 16.2

1. Let H be a subgroup of $\mathcal{A}(K)$ and let K_H be the set of elements of K that are fixed by every automorphism of H. Then K_H is a subfield of K.

2. Let F be a subfield of K and let $\mathcal{A}(K, F)$ be the set of all automorphisms of K which fix every element of F. Then $\mathcal{A}(K, F)$ is a subgroup of $\mathcal{A}(K)$.

The proof is left as an exercise.

DEFINITION

1. K_H is called the **fixed field** of H.
2. If $K_{\mathcal{A}(K, F)} = F$, we will call K a **normal extension** of F.

Note that in general all we can say is that $F \subseteq K_{\mathcal{A}(K, F)}$. We shall soon see the importance of normal extensions in Galois Theory.

THEOREMS 16.3

1. Let K be a finite extension of F. Then $[K : F]$ is finite and the order of $\mathcal{A}(K, F) \leq [K : F]$.

2. If K is a finite normal extension of F, then $[K : F] =$ the order of $\mathcal{A}(K, F)$.

3. If K is a finite normal extension of F and H is any subgroup of $\mathcal{A}(K, F)$, then $[K : K_H] =$ the order of H and $H = \mathcal{A}(K, K_H)$.

The proof of Theorem 16.3 requires several sophisticated arguments and will not be given here. Again there are many good texts dealing with the subject. We will also state the following theorem without proof.

THEOREM 16.4

Let K be an extension of F. Then K is a normal extension if and only if K is the splitting field of a polynomial $p(x)$ over F.

DEFINITION

Let F be any field, let $p(x)$ be a polynomial in $F[x]$, and let K be the splitting field of $p(x)$ over F. The automorphism group $\mathcal{A}(K, F)$ is called the **Galois group of $p(x)$ over F.**

A few observations are in order. First, if $p(x)$ is of degree n, then $p(x)$ has at most n roots. Also, if K is a splitting field of $p(x)$ over F, $[K : F] \leq n!$. (Why?)

Galois theory

Finally it is relatively easy to show that the Galois group of $p(x)$ acts as a permutation group on the roots of $p(x)$. We sketch the proof here.

THEOREM 16.5
Let $G = \mathscr{A}(K, F)$ be the Galois group of $p(x)$ over F. If the roots of $p(x)$ are $\{x_1, x_2, \ldots, x_n\}$, then G is isomorphic with a permutation group on $\{x_1, x_2, \ldots, x_n\}$.

Proof: To prove the theorem, it is sufficient to show (1) that every $\alpha \in G$ permutes the roots of $p(x)$ and then (2) no two elements of G have the same effect on every root of $p(x)$.

1. First let $p(x) = a_0 + a_1 x + \cdots + a_m x^m$, $a_i \in F$, $i = 0, 1, \ldots, m$, and let $r \in K$ be a root of $p(x)$ in K. Then $p(r) = 0$ so that $\alpha(p(r)) = 0$. But

$$\alpha(p(r)) = \alpha(a_0 + a_1 r + \cdots + a_m r^m)$$
$$= \alpha(a_0) + \alpha(a_1)\alpha(r) + \cdots + \alpha(a_m)\alpha(r^m)$$
$$= a_0 + a_1 \alpha(r) + \cdots + a_m \alpha(r)^m = 0$$

since $\alpha \in \mathscr{A}(K, F)$ and $a_i \in F$, $i = 0, 1, \ldots, m$. Hence $\alpha(r)$ is a root of $p(x)$ if r is. Thus, for $\alpha \in \mathscr{A}(K, F)$, α permutes the roots of $p(x)$.

2. Finally, let $\alpha, \beta \in G$. Recall that K is the minimal field containing F and all the roots of $p(x)$. Now if $\alpha(r) = \beta(r)$ for every root of $p(x)$, then $(\beta^{-1} \circ \alpha)(r) = r$ for every root of $p(x)$. Since $\alpha(a) = a = \beta(a)$ for every $a \in F$, $\beta^{-1} \circ \alpha$ acts as the identity map for every $a \in F$ and every root of $p(x)$. Therefore $\beta^{-1} \circ \alpha$ is the identity in $\mathscr{A}(K, F)$. Therefore $\alpha = \beta$ in $\mathscr{A}(K, F)$ as we had to show. ∎

The next sequence of results makes up the key theorem in Galois Theory and sets up all the main results in the subject.

THEOREM 16.6: FUNDAMENTAL THEOREM OF GALOIS THEORY

1. Let $p(x)$ be a polynomial over a field F, let K be the splitting field of $p(x)$ over F, and let $\mathscr{A}(K, F)$ be the Galois group of $p(x)$ over F. Then for any subfield L with $F \subseteq L \subseteq K$, $\mathscr{A}(K, L)$ is a subgroup of $\mathscr{A}(K, F)$.

2. Similarly, for any subgroup $H < \mathscr{A}(K, F)$, K_H, the fixed field under H, satisfies $F \subseteq K_H \subseteq K$.

3. The mapping $L \leftrightarrow \mathscr{A}(K, L)$ is a bijection between the subfields of K containing F and the subgroups of $\mathscr{A}(K, F)$ such that
a. $L = K_{\mathscr{A}(K, L)}$ for any subfield L with $F \subseteq L \subseteq K$,
b. $H = \mathscr{A}(K, K_H)$ for any subgroup $H < \mathscr{A}(K, F)$. Furthermore $[K : L] =$ the order of $\mathscr{A}(K, L)$ and $[L : F] =$ the index of $\mathscr{A}(K, L)$ in $\mathscr{A}(K, F)$.

4. Finally, L is a normal extension of F if and only if $\mathscr{A}(K, L)$ is a normal subgroup of $\mathscr{A}(K, F)$; and if L is normal, then $\mathscr{A}(L, F)$ is isomorphic to $\mathscr{A}(K, F)|\mathscr{A}(K, L)$.

We have either stated or proved most of the results needed to prove Theorem 16.6. Rather than present the proof here, we will continue to answer the question of solvability by radicals. (Parts of the proof of Theorem 16.6 will be done as exercises.)

First, recall that if $p(x)$ is of degree n over F and K is the splitting field of $p(x)$ over F, then $[K : F] \leq n!$. In fact, it can be shown that there always exists a polynomial $p(x)$ such that $[K : F] = n!$, and the Galois group $\mathscr{A}(K, F) = S_n$ in this case.

Next, for a certain very large class of fields, $p(x)$ is *solvable* by radicals if and only if its Galois group is a *solvable* group. Combining these remarks and recalling that S_n is not solvable if $n \geq 5$, we come to our desired conclusion: there cannot exist formulas for solving the general polynomial of degree ≥ 5 by radicals.

Hopefully, the inclusion of so many results without proof in this last chapter will be outweighed by the opportunity to see how so many of the major results of this book come together in the proof of a beautiful classical theorem. Hopefully this glimpse into one of the next steps in the study of algebra will spur you into continuing your study of the relevant portions of that subject. There are many books on the subject which follow naturally from this one and introduce the reader to some of the deeper and more elegant aspects of algebra. Several other classical problems are solved by the techniques of Field Theory and reading some of the other texts will give the student a glimpse into those problems as well.

EXERCISES
1. Prove Theorem 16.2.
2. Prove parts 1 and 2 of Theorem 16.6.
3. Prove as much as you can of the remainder of Theorem 16.6. You may use the stated results in this chapter as well as the results in the rest of the text.

INDEX

Abelian group, 16, 95
Algebraic element, 195
Algebraic system, 3, 13
Associates, 169
Associative law, 32
Associative system, 14, 138
Associativity
   of addition, 4
   of matrix addition, 6
   of multiplication, 4
Automorphism
   field, 211
   group, 81
   inner, 86
   ring, 152
Automorphisms, set of all, 82

Bijection, 76
Binary operation, 13

Cancellation laws
   in a group, 28
   in a ring, 141
Cartesian product, 53
Cauchy's Theorem, 109
Cauchy's Theorem for Abelian Groups, 97, 108
Cayley's Theorem, 115
Center of a group, 43, 99
Central element, 99
Centralizer, 101
Characteristic, 157
Characteristic zero, 157
Class
   conjugate, 104
   equivalence, 56

Class equation, 106
Closed under an operation, 37, 138
Coefficients of a polynomial, 171
Commutative law, 16
Composition factors, 129
Composition series, 129
Conjugacy, 104
Conjugate class, 104
Coset
   left, 62
   right, 62, 65
Cycle, 120
Cycle structure, 122
Cyclic group, 20, 45
Cyclic subgroup, 43, 44

Degree
   of an extension, 210
   of a polynomial, 171
Derivative of a polynomial, 201
Determinant of a matrix, 8
Dihedral group of the square, 23
Distributive laws, 138
Division algorithm, 46
Division in a ring, 164
Division ring, 160
Divisor, greatest common, 164
Domain
   Euclidean, 162
   integral, 158
   principal ideal, 163

Element
   algebraic, 195
   fixed, 211
   idempotent, 30

216   Index

Element (*cont.*)
  identity, 9
  inverse, 14
  prime, 166
  transcendental, 195
  unit, 166
Equivalence class, 56
Equivalence relation, 55
Euclidean domain, 162
Euclidean ring, 161, 162
Even permutation, 125
Extension, normal, 212
Extension field, 194
  degree of, 210
External direct product, 54

Field, 158, 183
  extension, 194
  fixed, 212
  Galois, 205
  prime, 194
  of quotients, 189
  splitting, 203
Fields, tower of, 208
Finite group, 95
Fixed element, 211
Fixed field, 212
Functions, 10
  addition of, 10
Fundamental Theorem of Arithmetic, 168
Fundamental Theorem of Galois Theory, 213

Galois field, 205
Galois group, 208, 212
Gaussian integers, 170
Generator, 45
Greatest common divisor, 164
Group, 15
  abelian, 16, 95
  of automorphisms, 83, 85
  cyclic, 20, 45, 61
  dihedral, 23
  finite, 95
  Galois, 208
  generated by g, 45
  of integers modulo n, 20
  Klein four, 19
  permutation, 24
  quotient, 69
  simple, 129
  solvable, 130
  symmetric, 112

Homomorphism
  group, 86, 87
  ring, 150
Homomorphism Theorem for Groups, 92

Ideal, 145
  left, 148
  principal, 163
  right, 148
  trivial, 149
Identity, additive, 3, 5
Identity element, 9, 14
Identity mapping, 81
Image, 88, 153
Index, 72
Injection, 76
Inner automorphism, 86
Integers, 5
Integral domain, 158
Inverse
  additive, 3, 15
  multiplicative, 4, 15
Inverse element, 14
Irreducible polynomial, 179
Isomorphic groups, 77
Isomorphic rings, 152
Isomorphism
  group, 75, 77
  ring, 152

Kernel, 89, 153
Klein four-group, 19

Lagrange's Theorem, 72
Law of exponents, 35
Left coset, 62
Left ideal, 148
Linearly dependent, 208
Linearly independent, 208

Mathematical induction, 32
Matrices
  addition of, 6
  additive identity of, 6
  additive inverse of, 6
  product of, 7
  2×3, 5
  2×2, 7
Matrix
  determinant of, 8
  inverse of, 9
  symmetric, 41
Minimal polynomial, 195

Noncommutativity of matrices, 9
Normal extension, 212
Normalizer, 102
Normal series, 129
  refinement of, 129
Normal subgroup, 66

Odd Permutation, 125
One-to-one correspondence, 76

One-to-one mapping, 76
Onto mapping, 76
Orbit, 119
Order of an element, 49, 51
  infinite, 49
Order of a group, 15, 51
  infinite, 15

Partition, 59
Permutation, 25, 111
  even, 125
  odd, 125
Permutation group, 24, 112
Polynomial over a ring, 171
  coefficients, 171
  degree, 171
  derivative of, 201
  irreducible, 179
  minimal, 195
  reducible, 179
  root of, 195
Polynomial rings, 170
Polynomials
  equality of, 172
  product of, 173
  sum of, 172
Prime element, 166
Prime field, 194
Principal ideal, 163
Principal ideal domain, 163
Product
  cartesian, 53
  external direct, 54

Quotient group, 69
Quotient ring, 148

Real numbers, 3, 4
Reducible polynomial, 179
Relation, 53, 54
  equivalence, 55
  reflexive, 55
  symmetric, 55
  transitive, 55
Right coset, 62, 65
Right ideal, 148
Ring, 138
  division, 160
  Euclidean, 161, 162
  polynomial, 170

quotient, 148
simple, 149
trivial, 155
with identity, 157
Root of a polynomial, 195
  multiple, 201
  of multiplicity n, 201
  repeated, 201

Series
  composition, 129
  normal, 129
  refinement of, 129
Simple group, 129
Simple ring, 149
Solvable by radicals, 208, 214
Solvable group, 130, 214
Splitting field, 203
Subfield, 192
Subgroup, 37
  cyclic, 43
  normal, 66
  proper, 42
Subring, 143
Surjection, 76
Sylow's Theorem, 110
Sylow's Theorem for Abelian Groups, 97
Symmetric group, 112
Symmetries, 22
System
  algebraic, 3, 13
  associative, 14

Table
  of the dihedral group of a square, 23
  of the integers modulo 5, 20, 60
  of the integers modulo 6, 137, 138
  of the Klein four-group, 19
  of the symmetric group on 3 elements, 26
Transcendental element, 195
Transposition, 123
Trivial ideal, 149
Trivial ring, 155

Unique Factorization Theorem, 169
Unit, 166

Well-defined operation, 68

Zero function, 11